Cultural China Series

Liu Junru

CHINESE FOOD

Adventures in the World of Eating

Translated by William W. Wang

China Intercontinental Press

图书在版编目（CIP）数据

中国饮食：英文/刘军茹著；(美) 威廉·王(William W.Wang) 译.
—北京：五洲传播出版社，2017.8

ISBN 978-7-5085-3762-7

Ⅰ. ①中… Ⅱ. ①刘… ②威… Ⅲ. ①饮食－文化－中国－英文 Ⅳ.
①TS971.2

中国版本图书馆CIP数据核字（2017）第196215号

中国饮食（英文）

著　　者：刘军茹
翻　　译：(美) 威廉·王(William W.Wang)
出 版 人：荆孝敏
责任编辑：王　峰
装帧设计：唐　妮
出版发行：五洲传播出版社
地　　址：北京市海淀区北三环中路31号生产力大楼B座6层
邮　　编：100088
发行电话：010-82005927，010-82007837
网　　址：http://www.cicc.org.cn，http://www.thatsbooks.com
印　　刷：北京圣彩虹科技有限公司
版　　次：2017年8月第1版第1次印刷
开　　本：710×1000mm 1/16
印　　张：9.75
字　　数：150千字
定　　价：96.00元

Contents

Preface

When it comes to food, the Chinese have a common saying, "The masses regard food as their heaven," which means that food is people's primal want. It should clearly justify the importance that "eating" holds in Chinese people's lives. Eating is not just meant to fill the stomach; having food at one's disposal, being able to consume a good amount of food, and knowing what and how to eat are all viewed as a good "fortune." Those who promote food culture often use the words of Chinese philosopher Confucius, "diet and love-making, all primal needs of every human being," finding an aspiring and positive thought basis for such an epicurean lifestyle. There's probably not another place in this world that has as great a variety of delicious

Rice fields by the Fuchun River (Photo by Miao Wang, provided by image library of Hong Kong *China Tourism*)

fare as China. If judging by the art and techniques of cooking, aside from France and Italy, perhaps the skills of chefs of no other country can gain recognition from the Chinese.

Extremely developed culinary techniques can make seemingly inedible ingredients, to foreign eyes, into dish after dish of delicious treats by the hands of Chinese chefs. The Chinese cookbook also contains quite an extensive list of foods, including just about anything edible with very few taboos. The Chinese, who see eating as a fortune and life as an art, not only created various kinds of regional food styles in its own vast lands, but have also spread Chinese food culture to far across the seas. Today, in this world where even the farthest corners can seem as close as one's back yard, Chinese food can be enjoyed in each and every metropolitan throughout the world.

Like many other countries with a vast territory, Chinese cuisines are differentiated largely by its northern and southern regional tastes. Although the best quality rice in China is grown in its northeastern regions, people in those regions, however, just as other northerners, prefer to eat pasta or pastry instead. In the north, classic

Vegetable sections in supermarkets that supply all kinds of fresh seasonal vegetable. (Photo by Huiming Shi , provided by Imagechina)

dishes include Beijing's lamb hotpot (fondue) and roast duck, and Shandong province's *Lu* style cuisines. In the south, the principal foods (foods that are the main source of carbohydrates and dietary fibers, e.g. bread and cereal in the west) are rice-based. A relatively greater variety of dishes are found in the south. There you can find the hot and heavily spiced *Chuan* (Sichuan) food, *Xiang* (Hunan) food, sweet and delicate *Huaiyang* food and the *Yue* (Cantonese) style which are mostly seafood and soups. Thus foreigners who have been to China are often pleasantly surprised by the great differences in taste and food types by region.

Having Chinese food not only indulges people's sense of taste, the sense of sight is also entertained. The Chinese culinary arts rely on the canon of "color (aesthetic beauty), aroma, and taste," missing any one element would not make a good dish. To make the food pleasing to the eye, usually the appropriate meat and non-meat ingredients are selected; it would include a single main ingredient and two or three secondary ingredients of different colors. Blue, green, red, yellow, white, black and brown sauce colors are to be mixed in the right combination. Through proper cooking techniques, aestheticism in food is achieved. "Aroma" is achieved by using the right spices, such as scallion, ginger, garlic, cooking wine, aniseed, cassia bark, black pepper, sesame oil, shiitake mushrooms and so on, to stimulate the appetite with the aroma from the cooked food. When preparing food, techniques such as fry, stir-fry, roast, steam, deep-fry, quick-fry, simmer and others are put to use, with the goal of preserving the natural taste and juices of the food. One can also add the right amounts of soy sauce, sugar, vinegar, spices, spicy pepper and

Scholar's Understanding of Diet before Qin Dynasty

The Pre-Qin period was a time of great turmoil and change in Chinese society, and it produced a number of great thinkers and ideas which were to have a deep and long-lasting influence. With regard to systematic reflection on drinking and eating, Mozi, Laozi and Confucius are typical in their different ways. Mozi had a very simple style of life. He advocated mutual assistance in society and active engagement in production, and thought that people should not eat unless they also toiled. He suggested that people should get only the food that their stomachs could hold and only the clothes that would cover their bodies. 'When it comes to food, there should be no more than suffices to replenish one's energy and fill the empty spaces; all that is required is to strengthen the body and satisfy the stomach.' He thought that people should live frugally and moderately and should serve society. Laozi drew attention explicitly to the importance of food and drink to self-cultivation. 'Those who would regulate the body and nourish the spirit must be sparing in their sleeping and resting and moderate in their eating and drinking.' He advocated purifying the heart and reducing one's desires, and knowing how to be content. His view of life emphasised spiritual cultivation and indifference to material things. Confucius integrated eating and drinking practices into the ritual system. His widely quoted saying 'There's no reason to reject the most carefully selected rice and the finest of chopped meat' is a call to ritual propriety not to luxury. These words had a great influence on the intellectuals of later times.

Dining environments with antique flavors and imitation imperial dishes of royal heritage closely bind culture and cuisine together. (Photo by Yu Shen, provided by Imagechina)

This is a piece of New Year's painting named *Abundance & Harvests in Successive Years*, which shows people's best wishes at the beginning of the New Year. (Collected by Shucun Wang)

other seasoning, making the dishes taste salty, sweet, sour, hot and much more. With tomatoes, turnip, cucumber and other sculptural vegetables to create elegant and intricate decorations to the plate, and the use of exquisite fine china for dining ware, Chinese cuisines really become a true art form complete with aesthetic beauty, wonderful aroma, and great taste.

Americans rely on calculating calories and cholesterol content from food to maintain good health and a shapely figure. The Japanese are into trying various health foods to preserve an everlasting youth. Different from both, the Chinese way of looking at health lies in its philosophy of "food and medicine sharing the same roots." The firm belief that food has healing powers and therapeutic effects has led to the introduction of many edible plants and herbs. And with the benefits of disease prevention and health preservation, they have become regular dishes in Chinese homes. At the same time, there is the pursuit of refinement in cooking.

The amount of food and mixing of ingredients is very essential, and it is recommended that meats and non-meats be used in combination. Whether making dishes or soups, foods with suited nutritional contents are put in combination so as to achieve the goal of balanced nutritional intake. And it is recommended to dine until the stomach is about 70% to 80% full, as this practice is passed down the generations as a secret to long life.

At the dinner table, the Chinese has their own set of manners and customs. When dining, the eater must be seated. When people of all ages and both sexes sit at the same table, the elderly must be seated with priority. One must eat food held with chopsticks; when having soup, a soupspoon must be used. There's also to be no noise when eating and so on. These etiquette have continued to this day, but the biggest change is none other than the fact that more and more Chinese have proactively given up the rule of "No talking when eating." Indeed, when dining with the Chinese, one would frequently encounter a dining environment full of chatting and noises. Many people who have their mouths full still intend to chat away. This phenomenon may be due to the reason that contemporary Chinese have come to consider dining as an important social opportunity. People need, at this time, to relax and talk about certain soothing and joyous topics to increase understanding between those sitting at the table.

In recent years, due to the accelerated development of industries and commerce, aside from traditional menu-ordered food services, Chinese fast foods have dawned onto the scene. And not only this, cuisines from every corner of the world have, one after another, made their grand appearances in all major cities in China; Italian pizza, French gourmet, Japanese sushi, American burgers, German beers, Brazilian barbeque, Indian curry, Swiss cheese and more. Anything one can think of can be found, a true all-inclusive list of dining choices. It justifies the saying "Eat in China" even more so.

The Origins of Food and Drink Culture

2,500 years ago, mountain residents in southern China invented the technique to reclaim rough mountain lands into fertile lands. They drew mountain spring water to irrigate and grew rice in terraces. The picture shows the terraces reclaimed by people of the Zhuang nationality in Guilin. (Photo by Guanghui Xie, provided by image library of Hong Kong *China Tourism*)

Tracing the Origins of Foods

There is a saying, the reason that great differences exist between eating habits of various regions of the world is the result of a multitude of factors, including limitations in ecological environment, the population volume, level of productivity and others. Most meat dishes are from areas where population density is relatively low and the soil is either not needed or unable to sustain agriculture. Reliance on meat has possibly stimulated economic activities of sharing and trade and. In comparison, a dietary habit of mainly grain, and plants' roots, stems, leaves and less meat is usually associated with an environment where supply cannot meet demand. The food supply in these places is more dependent on self-growing. However, dietary habits are not status quo, and with no classification as good or bad. But with migration of people on a global scale, dietary traditions that are once fixed to a region might be accepted and adopted by more and more people; and the original regional dietary habit evolves to contain more new elements. People could possibly see from the long-standing Chinese food culture the footprints of the common development of humankind.

China is one originating source of the world's agriculture. The Chinese have invented ways of irrigation at a very early time. Building canals and using sloped land to develop agriculture by irrigation, as well as other means of farming. As early as 5,400 BC, the Yellow River region already saw growth of foxtail millet (Setaria italica, also called foxtail bristlegrass, meaning the seed of broomcorn millet), and has already adopted the method of crop storage in underground caves. By 4,800 BC, areas along the Yangtze River have been planted with rice (with the distinction of sticky or non-sticky rice, the earliest "rice" pertains to the glutinous types of rice only). Since entering the agricultural age, the Chinese have formed a dietary composition with grains as the principal food and meats as supplement, and such tradition has continued to this day.

There exists an old piece of writing in China by the title of *Huangdi Neijing*. It describes the food composition of the Chinese as "The Five Grains as life support, the Five Fruits as complimentary aide, the Five Meats as added benefits, and the Five Vegetables as substantial fill." The grains, fruits, and vegetables are all plant foods. Grain crops in ancient times were referred to as "The Five Grains" or "The Six Grains," and usually consists of *shu* (broomcorn millet, sometimes referred to as "yellow rice," a small glutinous yellow grain), *ji* (what we call millet today, has the title of "Head of the Five Grains," *shu* and *ji* were the principal cereals of Northern China at the time), *mai* (including barley and wheat), *dou* (the general term for all pod-bearing crops, grows in wet lowland areas, and is the main source of protein for the Chinese), *ma* (refers to the edible type of hemp, was the principal food for farmers in ancient times), and *dao* (rice). The *shu* and *ji* are both indigenous to China, and were introduced to Europe in prehistoric times. Rice and wheat are not

Drying crops in the sun on rooftops is a more common tradition in the countryside of southern China. (Photo by Xiaoming Feng, provided by image library of Hong Kong *China Tourism*)

native to the north of China. It is generally thought that the origins of rice are to be found in South China, India and South East Asia. In the sites of the Chinese Neolithic Hemudu Culture (5000—3000 BC) , archaeologists have found the world's earliest evidence of rice cultivation, but in early times the growing of rice in the north of China was far from widespread, and rice counted as a precious grain. It was not until the Han Dynasty (206 BC to 220 AD), with steady improvements in irrigation and the opening up of the south, that rice gradually became a regular food everywhere, and even then for the most part white rice continued to be regarded as a comparatively expensive food. Wheat originated from Central or West Asia. Some time around the Neolithic period it was introduced to China from northwest, but cultivation started later than the growing of rice, and until the later years of the Zhou Dynasty (1046 to 256 BC) wheat was only for the aristocracy. Also, the sorghum is an indigenous Chinese crop as well, and was introduced to India and Persia (present day Iran) during the first century AD. During every Chinese New Year celebration, the Chinese use the idiom "Good Harvest of the Five Grains," which really means to bless the New Year with good harvest of all crops, so as to bring prosperity. This is enough to show that in a large country where "The masses regard food as their heaven," the production of crops has held enormous importance since olden days.

Experiences from cultivating land gave way for the Chinese to learn about many edible plants that are unknown to the West. And they have discovered that many of the human body's essential nutrients can be obtained from plants. The beans, rice, broomcorn millet, millet and other foods that the Chinese often eat are all rich in proteins, fatty acids and carbohydrates.

Foods made from grain come in many varieties and take on many forms. The northern Chinese's principal food was wheat. Therefore, most dishes on the dinner table are various types of pastry or pasta. Wheat flour is made into buns, pancakes, noodles, stuffed buns, dumplings, wonton and so on. On the other hand, in

Special racks to air-dry grain crops in the sun used in the villages of Guizhou Province. (Photo by Yinian Chen, provided by image library of Hong Kong *China Tourism*)

the southern part of China, the principal food is rice-based. Besides plain rice, there would be thin rice noodle, thick rice noodles, rice cakes, stuffed glutinous rice balls in soup and other types of pasta and pastry to be found everywhere. Rice spread from south to north, and with barley and wheat passing from west to east contributes significantly to the shaping of Chinese dietary habits.

Noodles after being air-dried can be stored for longer periods of time. (Photo by Michael Cherney, provided by Imagechina)

Bing, or Chinese pancakes, was one of the earliest forms of pastry. The earliest method of making *bing*, is to ground the grain to a powder, make into dough by adding water, then boil in water until cooked. In time, there has come to be steamed, baked, toasted, fried and other kinds of pancakes. *Bing* also has the most varieties among all dough-made foods. It comes in all sizes and thickness, some with stuffing. Even the stuffing comes in no less than several dozen varieties.

The non-stuffed pancakes are single or multi-layered. Those with good skills can make around a dozen layers in a pancake, each layer being as thin as paper. The sesame seed cake is the most popular baked pastry, and can be found in both the north and the south.

Noodles are also a type of traditional food made from flour. The earliest way of making noodles was nothing but to cook in boiling water or soup. It was only after the Song Dynasty (960—1279 AD), did there come to be meat or vegetarian pasta sauce. Noodles have a close correlation with Chinese festivities. In the north, there is the belief that "on the second day of the second month (lunar calendar), the dragon raises its head." So people have the custom of eating Dragon Whisker Noodles, to pray for good weather and harvest during the year. In the southern regions, on the first day of the lunar year, "New Year's Noodles" are to be had. In addition, Longevity Noodles are for celebrating birthdays. When a child reaches one month in age, together the family shall have "Soup Noodle Banquet." Though the art of noodle making may look

China is an important center of origin for citrus fruits in the world. Original wild orange types grow in many places such as Hunan, Sichuan, Guangxi, Yunnan, Jiangxi, Tibet and so on. (Photo by Yu Shen, provided by Imagechina)

simple, it is actually a complex task that requires many different skills, such as rolling, rubbing, stretching, kneading, curling, pressing, and slicing.

The Chinese at around the 3rd century AD, have mastered flour fermentation techniques by using the easily fermented rice soup as a catalyst. Later, bases were experienced to neutralize the fermentation process when making dough. The advent of the steam basket, the Chinese griddle and other cooking utensils, together with fermentation techniques, have helped to provide the endless possibilities of pasta dishes and pastry. The most common food made from flour, since the development of fermentation techniques, would be the mantou, or plain steamed bun.

Plain steamed rice is the most commonly encountered type of rice-food, and is the principal food of the southern Chinese. But more characteristic of traditional Chinese rice-foods is still zhou, or Chinese porridge (congee). Porridge has had thousands of years of history in China, and the way people eat porridge varies from

Stone millings are traditionally important tools to process crops. Before the 1950's, in many places, stone millings were also important dowries when girls get married. At present, mechanical processing technique is applied by more and more farmers in China. The picture shows a mill in a Shaanxi cave-house. (Photo by Xiaogang Shan, provided by image library of Hong Kong *China Tourism*)

Rice is a major source of the Chinese's principal food. From Daxing'an Mountain to areas along the Yangtze River, and from Yungui highlands to the foot of the Himalayas, wherever rice can be grown, it will appear in people's daily diets, religious celebrations and wedding banquets or in the paintings and songs. The planting of rice changes sceneries in its location. Almost 3 billion people in the world share the culture, tradition and the untapped potential of rice. The picture shows farmers transplanting rice seedlings in rice paddies in Hainan Province. (Photo by Yijun Xiong, provided by Imagechina)

region to region. There are also countless varieties of Chinese porridge, where just the basic ingredients are divided into six main groups, namely the grains, vegetables, fruits, flowers, herbs and meats. And the way of eating rice dressed with porridge has existed for quite some time.

Thirty years ago, rice and white flour were considered "fine foods," which most common folks are not able to have at every meal. Its counterpart, the "rough foods," were the real main dietary components of the Chinese, including corn, millet, sorghum, buckwheat, oats, yams, beans and so on.

Among all the "rough foods," soybeans gave the greatest

contribution. The earliest record of soybean planting was in the West Zhou Dynasty. Soybeans at the time were the food of farmers. It was not until the West Han Dynasty (26 BC – 25 AD) after the emergence of tofu, or bean curd, did soybean become acceptable to the bureaucrats and the literati class in Chinese society. To the present day, there are well over a hundred kinds of tofu and foods made from soybean milk. Chinese-grown soybeans and soybean products provide for an important source of vegetable proteins, and can be made into many premium sauces. Bean curd is placed somewhere between the category of principal and supplementary foods. It has since its creation evolved into many kinds of dishes, and has become typical Chinese home cooking. Different when compared to westerners' common use of butter and other animal oils, the Chinese mostly use vegetable oils such as soybean oil, vegetable seed oil, peanut oil, corn oil and so on.

Before the Qin Dynasty (221 BC) writings, fruits to make the most frequent appearances are peaches, plums and jujubes; and after those come pears, sour plums, apricots, hazelnuts, persimmons, melons, hawthorns, and mulberries; making rare appearances are Chinese wolfberries, Chinese crabapples, and cherries. Most of these fruits are indigenous to fruit trees in temperate zones of northern China, or have been introduced to China in prehistoric times. Of which, peaches, plums, jujubes and chestnuts were often used as ceremonial rituals. Peaches were exported from northwestern China by way of Central Asia to Persia; and from there, the peach found its way into Greece and other European countries. So it is unlike the common belief of the Europeans that peaches originated in Persia. Many other fruits that were indigenous to southern China, including tangerines, shaddock (pomelo), mandarin oranges, oranges, lichee, longan, Chinese crabapples, loquat, red bayberries and more, are gradually being consumed in broader areas.

During the transformation from a fishing-and-hunting society into agricultural society, meats were also once an important component of the supplementary diet of the Chinese, due to underdeveloped technology in the growing of vegetables. In the agricultural age, the Chinese considered cattle, sheep, and pig to be the three superior domesticated animals, called the three *sheng*, or sacrificial animals. When performing sacrificial rituals, the three animals were considered the best grade of all offerings. Horse, cattle, sheep, chicken, dog and pig together, were called the six *chu*, or domesticated animals. Under the influence of relatively high population density and limitations in the natural environment, as well as other factors, horses and cattle were most often regarded as principal assistants in agriculture, and not fed and raised as livestock for food. Therefore, all the way until the Song Dynasty, the Chinese considered beef a rare delicacy, whereas mutton was seen as a very common dish. Lamb (meat from a young sheep) was considered the superior grade of meat from a sheep. The character *mei* in the Chinese script, meaning beauty, is associated with eating mutton in its meaning and form. Pigs and Chickens were also some of the earliest animals to be domesticated and used as food. Due to the early development of poultry breeding, eggs are the most frequently consumed animal-related food for the Chinese. A common feature of the Chinese countryside is that families raised pigs (excluding believers of Islam), as pork is the most common meat in Chinese food. With the same attitude towards lamb, the ancient Chinese believed that meat from a piglet tastes better than that of a fully-grown pig. In China's past, dogs were animals that could be slaughtered at any time to be cooked as food. Though it is not as common as having pork and chicken, there were specialized professions in the area of dog butchers. The Chinese also invented the primitive egg incubator, breeding cell and many other poultry feeding devices.

Food for the Chinese since the pre-Qin period have been mainly grains, so meats were rare and cereals were abundant. With the advancement of vegetable growing techniques, vegetables were no longer the privileged enjoyment of the wealthy few. The list of vegetables that the Chinese eat is perhaps the biggest variety offering in the world. Common veggies include the Chinese cabbage, turnip and radish, eggplant, cucumber, peas, Chinese chive (leek), wax gourd, edible fungi, plant shoots, and various beans, as well as edible wild herbs grown in small quantities. Wild herbs are supplementary foods with the main purpose of helping people to swallow food. This forces culinary technique to constantly improve upon itself. The various vegetable roots, stems, and leaves could be eaten fresh or cooked, and could be dried for storage, or cured for making different kinds of appetizers. The goal is to offer as much variety in texture and taste as possible.

When compared to a dietary composition of excessive animal-based foods, many nutritional scientists believe that the Chinese inclination towards grains as principal food, with fish, meats, eggs, milk and vegetables being supplementary diet components, helps to provide for a balanced nutritional intake and more benefits to health, and is also in accordance with the global call for energy conservation and environmental protection.

Foods from Outside China

According to statistics, seventy to eighty thousand species of edible plants exist on the earth, among which are about 150 species that can be grown in large quantities. However, only 20 species of those are being used widely in farming today, but already make up for 90% of the world's total grain production. Domesticated animals and plant species are essentially the basis of global agricultural production. The fishing industry, which relies on wild animals as basic provision, outputs nearly 100 million tons of food annually for global consumption. Just as other countries of the world, the trade and spreading of foods, edible plants and animal species in China have been a non-stop process since ancient times. Not only did this broaden the range of the Chinese's food supply and made Chinese cuisines even more full of tasty dishes, but has also caused changes in Chinese dietary habits and put more life and variety into Chinese food culture.

Besides a small number of food species that were introduced to China by the pre-Qin period, a larger-scaled food trading and spreading happened over two thousand years ago during the tremendously prosperous and powerful West Han Dynasty. Grape, pomegranate, sesame, lima bean, walnut, cucumber, watermelon, muskmelon, carrot, fennel, celery, Chinese parsley (coriander) and other food species, which had its origins in the Xinjiang (Uyghur) region of China or West and Central Asia, made their way into central Han Chinese territory by way of the Silk Road.

And it was from that time, the Chinese and foreign cultures experienced more communication as the days went by. Many foods that were not indigenous to China began to appear on Chinese dining tables.

The corn, which has its roots in the Americas, was introduced to the Orient through Europe, Africa and West Asia. The potato, a cross between principal food and a vegetable, came to China via the southeast coast of China; at first it was only planted in Fujian and Zhejiang regions. Sunflower seeds made its way into China from America during the 17th century; 200 years later, cooking oil was extracted from it,

Although the history of Chinese eating hot peppers is only 300-some years, the custom of having hot foods is already quite popular. The picture shows a vendor selling red chili peppers. (Photo by Yunfeng Zheng, provided by image library of Hong Kong *China Tourism*)

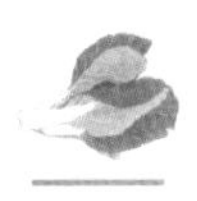

Eating a popsicle (Photo in 1957 in Beijing, provided by Xinhua News Agency photo department)

Coffee-making utensils. (Phot by Hao Wang, provided by Imagechina)

making the Chinese line-up of oils even more complete. The mung bean (gram), of the pod-bearing crops, has its roots in India, and was brought to China in the Northern Song Dynasty (960—1127 AD). Spinach, a kind of vegetable, came by way of Persia during the reign of Emperor Taizong (627—649 AD) of the Tang Dynasty. The eggplant, which was first found in India, along with the teachings of Buddhism, spread into China in the North and South Dynasties (420—589 AD). Many crops with unmistakably Chinese origins such as peanut, garlic, tomato, balsam pear, pea and other food types were replaced by premium foreign species.

Early fruits introduced to China mostly came from West Asia (e.g. grapes), Central Asia (e.g. early apples), the Mediterranean (e.g. olives), India (e.g. oranges), and Southeast Asia (e.g. coconuts, bananas). Other fruits such as pineapple, tomato, persimmon, strawberry, apple, durian, grapefruit and more, which have become the principal fruits for the Chinese, were imported from Southeast Asia, the Americas, or Australia/Oceania in modern times.

Hot pepper, already a popular type of spice for Chinese dishes, has only had about 300 years of history with the Chinese. Historical records show that hot peppers came to China by sea from Peru and Mexico during the late-Ming Dynasty (1368—1644 AD). Sugar, the main source of sweetness in cooking, saw its production in China after Emperor Taizong's ambassadors to Central Asia during the Tang Dynasty, learned sugar-making skills. What the Chinese see as high-class food, namely the shark's fin and bird's nest, were introduced to China from Southeast Asia in the 14th century. Starting in the Qing Dynasty (1616—1911 AD), they have become lavish foods for the wealthy only. With the widespread influence of Western cultures, exotic beverages such as coffee, soda, fruit juices as well as all kinds of alcohol drinks are no longer a rarity in Chinese eyes.

In terms of dishes, the earliest foreign recipes were introduced to China in the Tang Dynasty. As frequent trade between China and other countries flourished, the Arabs, who brought their Muslim foods, made great contributions to diversifying the Chinese dietary customs and adding to the already plentiful selection of Chinese culinary techniques. In near-modern times, Western foods appeared in China. Not only can all types of Western restaurants be found around many commercial ports, Chinese and Western food even fused together to create a new style of gourmet technique. This is most exemplified in the *Yue* (Cantonese) style of Chinese foods.

In recent years, as Sino-foreign economic and cultural exchange became more intimate, the importation of premium animal and plant species from foreign countries has already become a crucial part of the Chinese import business. More and more foreign foods have found their ways into the home of common Chinese families. However, the Chinese government, just as governments of other countries, is beginning to see the large quantities of imported, or invading foreign species as a threat to domestic biological varieties. Laws and policies on protecting national ecological security are been drafted and implemented.

Fine Ware for Eating

Humans evolved from the primitives, who plucked the hairs and feathers from animals and drank blood, into intelligent and skillful beings that can make today's gourmet foods. Gone were the days of seizing food with bare hands, people now dine with chopsticks, knives, forks and spoons. Apparently, changes in the ways of eating and dining utensils can reflect the path of human evolution, from a primitive state to modern men. The cooking and dining utensils of the Chinese have an inseparable connection with their culinary techniques and dietary habits. Today, people can learn about history through artifacts and a written language that were passed down through the generations. Chinese dining ware has gone through changes in material, from stone and pottery to bronze, iron and other metals. The one form of "made in China" product that is well known throughout the world is porcelain, or fine china. As productivity levels heightened, dining utensils not only underwent changes in material and craftsmanship, but also a typical change from large to small, rough to delicate, and thick to thin.

Since the Neolithic Age, pottery *dings* have been used as primary cooking utensils. By late Xia Dynasty (approximately 18th century BC—16th century BC), other than still performing as a cooking vessel, bronze *dings* were also used to hold dishes of meats in times of sacrificial rites. This ding is one of the earliest bronze *dings* still existing in China, with a height of 18.5cm and a diameter at the opening of 16.1cm.

The earliest cooking utensils included earthenware *ding, li, huo, zeng, yan* and more. Later came more elegant and larger successors to these utensils with the same names, but made from bronze and iron. Some of these cooking utensils doubled as vessels for food, such as the *ding* that was used to both cook and hold meat. Usually large in size, the *ding* is usually round in shape and has three pedestals for support; certain ones are square with four pedestals. Between the pedestals, firewood and fuel can be placed for direct burning

Gui is a kind of ancient vessel for millets and rice. This bronze gui was made at around 11th century BC with a height of 14.7cm and a diameter at the opening of 18.4cm.

and heating. On either side of the upper exterior of the *ding* is a handle for easy carrying. In the Bronze Age, the function of the *ding* changed as some were used as important tools in sacrificial rites. *Li* is used for cooking porridge (congee). It is similar in shape to the *ding* but smaller in size. Its three pedestals are hollowed and connect to the belly. The food in the hollowed legs therefore can be heated and cooked more quickly. *Huo* is specially used for cooking meats, and is more advanced than the *ding*. It has a round belly but no feet, more akin to the "wok," which came at a later time. The *zeng* is used for steaming food. Its mouth folds outward and has handles. The bottom is flat with many apertures for the passage of steam. Some *zengs* have no bottom, but instead has a grating underneath. When in use, the *zeng* is placed over the *li*, a cooking tripod filled with water. What merged the *zeng* and *li* together is the *yan*. The Chinese have had earthenware *zeng* since the late Neolithic Age. After the Shang Dynasty (around 17th to 11th century BC), there appeared *zengs* made of bronze.

Yan is a kind of ancient cooking utensil. This bronze yan was made at around 13th century BC—11th century BC with a height of 45.4cm and a diameter at the opening of 25.5cm.

Food containers had their divisions of responsibilities as well. Among remaining artifacts, besides the plates and bowls, which differ little in function from today's versions, there are also the *gui*, *fu*, *dou*, *dan*, *bei* and more. The *gui* is very much like a large bowl, with a round mouth and large belly. On the underside is a round or square base. Some has two or four handles at the upper outer rim. This kind of vessel was initially used to store grain, and was later used as a dining utensil, as well as a ritualistic tool. The ancient Chinese usually first fill rice from the *zeng* into the *gui* before eating. *Fu's* function is similar to that of the *gui* and its form is close to that of the later high-legged plate. But most *fus* have a lid. The difference between *dou* and *fu* is that a *dou* has handles at their bottoms. Earthenware *dou* surfaced during the late Neolithic Age. After the Shang Dynasty, there were wooden painted *dou* and bronze *dou*. The *dou* is not just a dining utensil, but a tool for measurement as well (in ancient times, 4 sheng make up one *dou*). *Dan* is a container for rice, made of bamboo or straw. The *bei*, or cup, is not so different from today's cup in form or function, mainly for carrying soup. Regardless of having meat or rice, the bi is used. The *bi* that is used to get pieces of meat from the *huo* is larger in size than the smaller *bi* used to obtain rice from the *zeng*. The *bi's* function is the same as the modern-day spoon which succeeded it.

China has a long history of winemaking. Countless numbers of wine vessels from the Shang Dynasty were uncovered. From these, it can be determined that drinking alcohol was high fashion at the time. The *zun* (full round belly, protruding extended mouth, long neck, and spiral pedestals at the bottom, comes in many different designs and types of production process and material. The most popular during the Shang Dynasty was the "bird and beast" design), *hu*, (long-necked and small opening, deep-bellied with round base, some with overhanging handle), *you* (elliptic opening, deep-bellied with round base, has a lid and overhanging handle), *lei* (some round and some square, openings vary in size, short neck with square shoulders, deep bellied, with curling feet or round base, and has a lid), *fou* (earthenware) and the like are wine vessels, *jue* (deep-bellied, tri-pedestal, can be heated on top of fire, protruding grooves at the top for easy pouring), *gu* (the most common wine-drinking utensil, more often used in combination with

A jade bowl in the shape of lotus leaf, made in the Ming Dynasty (1368—1644) with a height of 5.3cm and a diameter at the opening of 9.4cm.

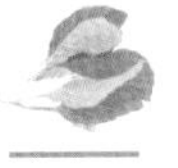

jue, which is bigger, opening is shaped like a bugle, long neck, thin waist, tall curling feet), *zhi* (similar to the *zun* in shape but smaller, some with lids), jia (round mouth and belly, three pedestals with short handles, used to warm up alcohol), *gong* (oval belly, has an outer edge for flow of wine, short handles, bottom has curling feet, lid is in the shape of the head of a beast, some has an entire body like a wild animal, with small spoon as accessory), *bei* (cup), *zhen* (shallow and small cup) is to drink from. Wine is stored in large containers such as the *lei*. When serving, wine is poured into *hu* or *zun*, and placed at the side of the seat and table, then poured into *jue*, *gu*, or *zhi* for drinking.

Great inventions such as gunpowder, compass, movable-type printing and many more appeared, bearing witness to the advancement of Chinese science and technology. At the same time, Chinese porcelain-making crafts reached unprecedented heights in the Song Dynasty, as the making of celadon porcelain, white porcelain, black porcelain, overglaze or underglaze enameling all experienced great improvements. There emerged much more creativity in modeling, patterns and decorative illustrations, and enameling. Many fine and rare porcelain pieces that are famed today in both China and the West were created during this time period. Exquisite porcelain vessels for food and wine, together with the Chinese tradition in "pursuit of refinement" in food, became the very much-treasured heritage of Chinese food culture that makes the Chinese so proud.

Speaking of the prominent characteristics of Chinese food culture, the chopsticks that the Chinese use for dining come to mind. The three main types of human dining tools are the fingers, fork and chopsticks. Seizing food with the fingers is a custom mainly practiced in Africa, the Middle East, Indonesia and the Indian subcontinent. The Europeans and North Americans use forks for dining. Other peoples, like the Chinese, who use chopsticks for dining, include Japanese, Vietnamese, and the North and South Koreans. Due to the influence of overseas Chinese, using chopsticks is also rapidly becoming a prevailing trend for people in Malaysia, Singapore and Southeast Asia.

There is a legend about the origin of the chopstick. In ages past, during the time of legendary sage kings, Yao and Shun, floods overrun the land causing catastrophes. Dayu received orders to prevent floods by water control. One day, Dayu set up a cauldron boiling meat inside. Meat after being cooked from boiling waters usually must be cooled before picking up by hand, but Dayu did not want to waste any time. He chopped off two twigs from a tree and used them to clip pieces of meat from inside the boiling soup. His men saw that with the twigs he was able to eat the meat without burning his hands and kept free of grease. So they imitated his doing one by one. Gradually, the rudimental form of the chopstick was established. The legendary story of Dayu inventing the chopsticks was a way the ancients honored the hero. Functionally speaking, chopsticks were introduced as a convenient method of picking up food which could burn the hand when hot. The Han period book *Shuowen Jiezi* (Explaining Graphs and Analysing Characters) explains the character *Zhu* as 'to grip from both sides and raise', and "to grip with wood". It is evident that the ancient Chinese people firstly used thin strips or bamboo as the eating implements.

Historical materials clearly record that in the Shang Dynasty, some 3000 years past from today, the Chinese have already begun the use of chopsticks when dining. The oldest pair of chopsticks preserved today is made of bronze, uncovered at the Yin ruins (ruins of late-Shang Dynasty capital, located in present-day Anyang of Henan Province. It is the earliest capital city in Chinese history with a confirmed location. Oracle bone inscription used for divination was discovered there in 1899. Large-scale archeological excavations at the site began in 1928). Upon reaching the Han Dynasty, chopsticks were already widely used by the Chinese. In the history of human civilization, the use of chopsticks by the Chinese is a scientific invention of which they can rightly be proud and prestigious. The famous overseas Chinese physicist Li Zhengdao once made this assessment of chopsticks: 'Such a simple pair of things, but they make use of the lever principle in physics in a truly remarkable way. Chopsticks are an extension of the human fingers and can do everything that the fingers can do, yet they are impervious to extreme heat or cold—really brilliant'.

The chopstick, great invention of the Chinese, is probably very much associated with mass consumption of the roots, stems, and leaves of vegetables in the Chinese diet. Long before invention of the chopstick, the main utensils for eating meat were *bi* (spoon), *dao* (knife) and *zu* (chopping block). Meat would be cut with the knife and served by hand. So even the ancient Chinese had the habit of washing hands before meals. Chopsticks have yet more importance as it influenced the course of development of Chinese dishes and dietary habits. For example, having foods such as lamb hotpot, long noodles, and bean-starch noodles just becomes that much more fun and convenient when the chopstick joined in.

Compared to knives and forks, chopsticks seem more difficult to

It is difficult to find a better replacement for Chopsticks when having certain Chinese cuisine and snacks, such as when enjoying Lamb Hotpot.

handle. The two thin sticks have no direct point of contact. Rather, with the thumb, index and middle fingers doing the work, the sticks can perform multiple feats including raise, stir, nip, mix, and scrabble. And it can precisely pick up any food except for soup, stew and other kinds of liquid foods. Specific studies show that when using chopsticks to clip foods, it involves more than 80 joints and 50 pieces of muscles in the body, from shoulders to the arms to the wrist and fingers. Using chopsticks can make a person quick-witted and dexterous. Many westerners praise the use of chopsticks as the creation of an art form. Some even think that the Chinese's excellent skills in table tennis should give credit to the chopstick.

However, chopsticks still have its weakness when compared to knives and forks, as well as to eating from the hand. When it comes to round and slippery foods such as stuffed glutinous rice balls, meatballs or pigeon eggs, one's skills of using chopsticks are put to the test. Those with less than average mastery of the tool can result in embarrassing moments.

Westerners have a real cultured way of dining, usually holding the knife in the right hand, fork in the left, and eating ambidextrously. The Chinese also has its own set of rules when eating. Chopsticks are for rice and spoons are for soup, but only one hand can be used at a time, unlike the west where both hands are used simultaneously from left and right. In addition, there are even more proprieties when eating with the chopstick. It is usually accepted that chopsticks should be held in the right hand. In olden days, training was conducted for the right-hand usage of chopsticks. When finished with the meal, chopsticks must be securely bridged on top, at the middle of, the empty bowl. If for a temporary recess in the middle of a banquet, chopsticks can be laid on the table close to the bowl and should not be placed upright in the bowl. This is because only bowls holding sacrificial offerings are to have a pair of vertically implanted chopsticks. It is also not allowed to aimlessly stir around amidst foods or to poke at things with chopsticks. When two people try to clip up foods, the pairs

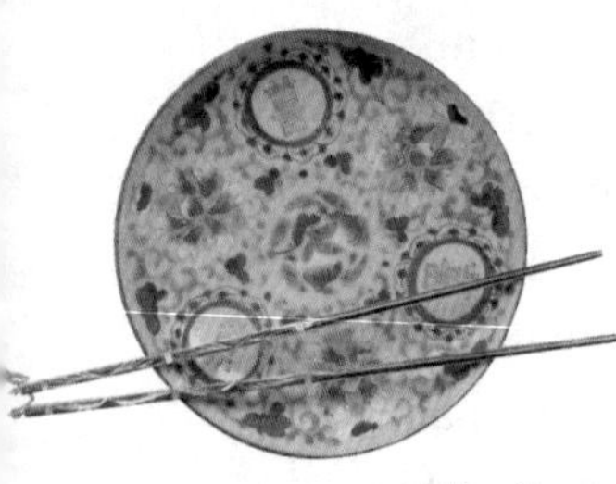

Many people have collecting chopsticks as a favorite hobby, but these collector's items are rarely used to eat daily meals with.

of chopsticks cannot cross. One should never knock on an empty bowl with chopsticks. One also cannot use two chopsticks of different lengths or use only one chopstick. Chopsticks cannot be used in place of toothpicks, etc.

As a daily appliance of the Chinese, its not unusual to find chopsticks made with a wide range of materials including bamboo, wood, gold, silver, iron, jade, ivory and rhino horn. Former kings and emperors of China usually dine with chopsticks of silver, as it has the peculiar property of reacting to poisonous chemicals by turning black, thus guaranteeing safety of the food.

Chopsticks are not only the most loyal "attendants" on Chinese dining tables, but also a cultural folk craft worthy of collecting. Therefore, many areas in China produce "brand-name chopsticks" that are made of exquisite materials through special crafting process. The unique artistic value that chopsticks hold has won the hearts of domestic and international tourists and collectors alike. Shanghai collector, Mr. Ling Lan, had a keen vision as he set up China's first family museum specializing in the collection of chopsticks. There, on display are over 1,200 pairs of chopsticks of over 800 types, all in extraordinary splendor for the viewing pleasure of visitors. Among the collection are hotel-used chopsticks; chopsticks from specific tourist spots; chopsticks used for dying of cloth in the countryside; Mongolian chopstick-dance props; metal chopsticks used as weapons in ancient armies; bird-raising chopsticks and more, the list goes on and on. In Indonesia, an elderly overseas Chinese has over 908 types of chopsticks in his collection, among which is a pair of golden chopsticks that was used by a former imperial Chinese concubine.

Food and Drink Traditions

The Good Cheer of Eating Together

China not only has a wide variety of cuisines and exotic fare in all its regions, even ordinary homemade cooking for three meals a day can provide for plentiful unique recipes. The Chinese stress the aesthetics of food, the refinement of dining ware, and the elegance of dining environment, so having food is a daily enjoyment. Eating, as a branch of learning and art form, not only gave birth to rich and excellent culinary techniques, also reflects the Chinese's content and joyful nature.

The Chinese have had a regular dining discipline since long ago. First it was a two-meals-a-day practice. The first meal, called zhao shi (morning food), is usually had around nine-o'clock in the morning. The second meal, bu shi, is had around four in the afternoon. The Chinese sage Confucius says that "bu shi bu shi," which translates to "meals are not to be had if it is not the appropriate time," meaning to emphasize the punctuality of meals. At around the Han Dynasty (206 BC—220 AD), with better development of agriculture, people of every nationality group and region slowly began to adopt the "breakfast, lunch and dinner" practice. Only their dinners were had much earlier than modern men, as they believed "work starts with the break of dawn and rest is to be taken when the sun goes down." Three meals of the day must be prepared and eaten fresh, a way of showing the Chinese's crave and love for food. In recent years, the

Deep-fried Twisted Dough Sticks and soybean milk are a favorite kind of breakfast for the Chinese. (Photo by Jianming Liu, provided by image library of Hong Kong *China Tourism*)

Many urban dwellers like to eat at breakfast stands on the street. (Photo by Xiaogang Shan, provided by image library of Hong Kong *China Tourism*)

Mantou, or steamed bun, is one of the Chinese's principal foods. Many families living in northern China can make it at home, but some people would rather buy ready-made steamed buns on the streets for convenience. (Photo by Yuanhao Ma, provided by image library of Hong Kong *China Tourism*)

All family members dining together around the table reflects Chinese ethics that cherish family value. (Photo taken in 1950, provided by Xinhua News Agency photo department)

Soup-filled buns, dumplings and rice dumplings all cannot be done without steam boxes. (Photo by Yankang Yang, provided by image library of Hong Kong *China Tourism*)

pace of life for urban Chinese are getting faster and faster. Dining out is becoming more and more common, especially for lunch. Most office professionals dine at nearby restaurants, or in cafeterias of schools and work units. As for dinner, female heads of households are usually very attentive in its preparation.

In comparison with the Western way of individually served foods, the way of shared dining is seen as a distinguishable characteristic of the Chinese. For the Chinese, no matter if it is dining with family or with friends and associates, people usually sit around the table and eat from the same served dish and the same bowl of soup. But this was not always the case, as ancient Chinese practiced individually served foods for quite some time before the switchover.

The earliest cooking and dining utensils were mostly earthenware placed on the ground. Later, supporting tools were invented, such as low, wooden tables. In the Shang Dynasty (around 17th to 11th centuries BC) oracle bone writings, one can see the character "su." Its image is that of banquet seat (mat) with people sitting on it. The character "xi," is the pictograph of mat, shows that Chinese men

A capable homemaker preparing holiday dinner for her familly. (Photo taken in 1980, provided by Xinhua News Agency photo department)

at the time sat low on the ground. Most seating was rectangular or square in shape with the distinction of different sizes and lengths. The longer ones could seat several people while the shorter ones can seat at most two. The square ones were called du zuo (single seat), used by the elderly or people with high status. According to the needs, single or multi-tiered small mats can be set up. A person's status can be observed from the number of people he or she is sitting next to. At the dining seats, a strict set of proprieties exists. Elders and younglings, or the noble and the vulgar, may not sit together. There exists factual occasions after someone has broken the proprieties, where the person who was dishonored wielded his sword to cut the seat into halves so as to end the inappropriate and humiliating seating situation. Corresponding to the seating customs, the Chinese also had a small table for each eater. The practice of individual seating and servings continued until the later years of the Han Dynasty (206 BC—220 AD). In Chengdu, Sichuan province,

Scenes of women and children dining in an official's home in southern China during the late 19th century.

uncovered from an Eastern Han Dynasty (25 BC—220 AD) tomb site were brick paintings with scenes of banqueting. People were shown to be sitting in groups of two or three, while tables lay in front. These paintings depict the lives of people during those times.

The ancient Chinese's dining habit cannot be explained independently of their dining utensils. In the Tang Dynasty (618—907 AD), the former dining habits changed suddenly. Tall and long dining tables and chairs appeared. From the wall paintings at Dunhuang's 473 caves, we can see illustrations depicting scenes inside a tent, where a long table sits, and tablecloth drapes from all four sides. We can see spoons, chopsticks, cups, plates and other dining utensils and vessels depicted on the table. Long seats are placed alongside the sides of the table, on them sits a number of men and women. Using high tables and large seats for dining gradually replaced the practice of seating on the ground. Sitting on round stools or high chairs around a table in a natural posture, while sharing a table full of tasty food, is the way the Chinese eat today; it is the most characteristic of the Chinese's dining habits. It is safe to say that the emergence of shared dining and related customs is inseparable from and based on the changes in dining utensils and seating.

The idea of having families and friends enjoying great food at the table, for the Chinese, is full of warmth and harmonious atmosphere. This may be related to the close attention the Chinese pay to blood relationships and kinship. On another level, traditional Chinese culture focuses on "he," or harmony. When dining together, it is an important way for the Chinese to have increased interpersonal understanding and communication. This may also be the reason behind the Chinese's preference to chat vigorously at dining banquets. Epicures are concerned for individually served foods impacting the preservation of culinary aesthetics. For example, when a whole steamed fish, with great color, taste and aroma, is to be divided into individual servings, how shall it be divided? Who gets the head and who gets the tail?

This certainly is a dilemma. It is no wonder that some epicure worry about China reverting to individual servings of food, afraid that it may result in a setback of a glorious culinary tradition, losing its certain unique advantages.

In the face of the threat of SARS in 2003, whether or not to have individual food servings was an unprecedented topic actively debated by the government and the general public. For a time, forced implementation of individual servings took hold in restaurants. However, with the disease slowly being put under control, people regained their tradition of shared dining. In actuality, with the increased popularity of buffets, as well as Chinese or Western fast foods, individually served foods have righteously entered the daily lives of urban Chinese. And with even more international communications, some upscale banquets have universally adopted the practice of individual servings but with an atmosphere of shared dining.

Both for communal meals and individual eating, the Chinese cuisine is particular about matching vegetable and non-vegetable dishes, and there is a proper order for serving cold and hot dishes and savoury and sweet. At formal banquets the sequence in which dishes are ordered and brought to the table is fully thought out. In the past, the more fastidious restaurants had a set system of courses. To take as an example the standard grand banquet in north China, generally four cold dishes were served first, mostly containing meat or seafood. These were an accompaniment to the initial rounds of wine, and keen drinkers often had eight such dishes instead of four. Next came four hot dishes from the wok, slightly greater in quantity than the cold course, with a preference for whatever was fresh and in season, not too oily, but light and pleasing to the palate. Then came four bowls of braised hot dishes, with a lot of sauce or gravy, which would remain hot and stimulate the appetite. Only after that were the main dishes properly brought in; these often contained delicacies of land or sea, and not only were the flavours wonderful, but the diners also marvelled at the skill of the chef; the vessels these main dishes were served in were out of the ordinary, in former times often being vast bowls containing enormous quantities. After the main dishes had been served it was time for sweet dishes, sweetmeats, congee or rice. In the end came soup and fruit. If the meal was Guangdong style the soup opened the meal. Nowadays, this set order of courses is only followed in formal banquets.

The Tray Level with the Eyebrows (The Behavior Between Husband and Wife)
As social rituals developed, the Chinese began to use small food trays to serve food in individual portions. After at least 3000 years of development finally in Tang times 'assembled meals' in the modern sense came into existence. On the subject of dishes served one by one, there is saying with a story related to it. In the 'Biographies of Recluses' in the *History of the Later Han* we read that the recluse Liang Hong, after being employed in the *Taixue* (National University), gave up his public career and returned home to marry Meng Guang. Taking her with him he moved to Wu Commandery (now Suzhou), where he earned his living as a hired labourer. When he came back from work each day, Meng Guang always had a meal ready for him, and she lifted the food tray in front of her forehead and served him with deep respect. Meng Guang's habit of raising the tray to the level of her eyebrows has become an often repeated way of alluding to the mutual respect and affection of husband and wife.

The Taste of Home Cooking

The three daily meals enjoyed by Chinese families are what we call "common home gourmet." Most ingredients found in home-cooked meals are taken from ordinary grocery and spice list. And the only principle that it abides by is good flavor. The so-called "common home style" also means that it is flexible and ever-changing, full of varieties and does not stick to just one form of cooking. The Chinese, who pay great attention to food, will not settle for bland and identical taste everyday. Under the tenet of a simple and non-luxurious life, cooking homemade dishes is certainly no easy task, as the food not only must entertain the taste buds of the family members, but must also be constantly changing in variety and combination. In general, home meals do not differentiate between "regional styles." However, due to the fact that China has a vast expanse of territories, with products and living habits different in each area, it objectively creates for the situation where home-style cooking tastes different in each and every home.

It is common belief that dinner is the one meal that the Chinese take most seriously, whereas breakfast is the simplest. At a breakfast table of the Chinese,

the most common food is the stuffed or plain steamed buns with a bowl of porridge (congee) and a dish of pickled veggies; we could also see wonton, hot soup noodle, rice and stir-fried dishes. Though the "deep-fried twisted dough stick" and soybean milk are standard breakfast items, few families make them at homes, as they are usually purchased from breakfast shops. Milk, oatmeal, or egg and ham sandwiches are no longer rare and fancy in the eyes of the urban population. Eggs and bean curd are the general source of protein in breakfast and are easy to prepare. For lunch and dinner, aside from rice and pasta, there are also stir-fries, soups and porridge for complement. The preparation of homemade foods is usually the responsibility of the female heads of households. But in families with double income, where both the man and woman earn a living, it is not uncommon for a man to make the meal.

Different from the West, the majority Han Chinese and most Chinese minority nationalities have little dairy beverage each day. But for the northwestern minority nationalities, dairy products are an important component of daily diet.

A street-side vegetable market in the 1970's in Shanghai. (Photo taken in 1978, provided by Xinhua News Agency photo department)

In areas where pasta and pastry are the principal food components, homemakers can usually use wheat flour, corn flour, sorghum flour, soy flour, buckwheat flour or naked oatmeal flour to make a wide variety delicious treats. According to different taste preferences, pasta dishes can be stir-fried, fried, stewed, steamed, braised, simmered and so on. When having a bowl of pasta, jiaotou, or pasta sauce, is a top priority, and usually comes in the form of fried bean sauce (usu. with minced meat), soy gravy, dipping, soup stock and so on. Second priority to eating pasta are the shredded vegetables mixings, the types of vegetables vary during the seasons. At the birthplace of Chinese pasta, Shanxi Province, there are at least 280 types of pasta in the cookbooks.

Sharing Good Food with Others

There is a tradition in China of respect for the old, and in everyday life almost all parents will teach their young children, either through games or personal example, to share delicacies with parents or elders as a token of respect and love. Likewise, people retell old stories which describe how the people of past times in lean years kept hard-earned food for their mothers even though they had not eaten enough themselves, thereby perpetuating respect for the old.

For homes with rice as principal food, nothing is more common than a pot full of steamy, savory rice. But day after day, this becomes rather monotonous. So people spent much time in coming up with different ways of cooking and combinations. Steam, boil, stir-fry, roast, deep-fry and simmer, different ways of cooking bring out drastically different texture and taste in rice. In daily life, the Chinese usually would not and need not have excessive meat dishes. More often, inexpensive vegetables with good value are preferred. Turnips or radishes, green vegetables and bean curd are almost indispensable from each Chinese household. Green turnip, white turnip, radish and carrot are available throughout the country all year round. These veggies can be eaten raw, boiled, stir-fried, pickled and so on. "Green vegetables" include Chinese cabbage, spinach, rape, celery, Chinese chive (leek), mustard and more, where we eat their leaves and stems. Common ways of cooking green vegetables

Pasta and pastry come in many varieties and shapes like artwork. The picture shows a Shandong lady making a meal for her family. (Photo in 1980, provided by Xinhua News Agency photo department)

Making dried vegetables is a tradition for farmers in many places, which still exits nowadays. (Photo in 1961, provided by Xinhua News Agency photo department)

is none other than cold dish with dressing, cooked or stir-fried, and boiled or stewed. If one dislikes certain vegetables for they do not facilitate the eating of rice very well, small amounts of meat or eggs can be mixed in for stir-frying. For bean curd, the most common and simplest way to prepare is to serve cold with sauce or boil with water then dress in soy sauce, sesame oil sauce or other sauces. *Tofu* that is deep-fried and then baked with sauce or stewed with vegetables is also very common. In all the Chinese restaurants around the world, *Mapo Tofu*, meaning numb-hot bean curd, can be found. It is prepared simply by placing diced bean curd into pre-cooked minced meat, and after fully boiled, add some hot sauce and Chinese prickly ash powder, and there you have a fully prepared dish. Aside from *tofu*, other kinds of food made from beans, which belong to the same family as bean curd, are also common dishes during all seasons of the year.

A plate of green vegetables and a plate of fish. (Photo by Roy Dang, provided by Imagechina)

Eggs are a major source of animal protein for the Chinese. (Photographed in 1959, provided by Xinhua News Agency photo department)

The Chinese usually classify meat dishes into four categories, which are "chicken, duck, fish and meat." But with developments in modern breeding industries, pork has replaced chicken, and became the most common meat for most Chinese nationalities, with Han Chinese especially. In the past, it was because pork is hard to come by, now because it can be enjoyed very often; ways of cooking for pork are the most numerous among all the meats. These include stir-fried, simmered, white-cut with sauce, twice-cooked, steamed with sauce and turned upside-down, steamed with ground rice, boiled in hot oil and so on. Most families can make these dishes. For home-cooked pork, many people like to use starch, so pork stir-fry feels even more tender.

The Chinese have quite a long history of breeding

chicken. Since ancient times, the Chinese have considered Chicken to be of great taste and regard chicken soup as a great tonic drink. Chicken can be steamed or stewed in clear soup, simmered with soy sauce, white-cut with dipping, or stewed in yellow wine, with no less than a dozen ways of preparation. Just the recipes for chicken can be compiled into thick books. Compared with chicken, ducks have a much higher price tag in northern China. Common northern families seldom cook duck though, as the famed Beijing (Peking) Roast Duck has to be enjoyed in special restaurants. The place most skilled in making duck dishes is the Jiangsu-Zhejiang area. There, Salty Watered Duck and Laobao Duck are not just treats of recognition of restaurants; many homemakers can make the dishes well too. There are many ways to cook fish as well. Most times, fresh fish is steamed or simmered in clear soup. Fish a little lesser in freshness would be braised, or one can add sugar and vinegar for Sweet and Sour fish. Beef and mutton are the principal foods of western Chinese minority nationalities. Most common

People generally believe in the nutritional value of Chicken soup. (Photo by Jianming Liu, provided by image library of Hong Kong *China Tourism*)

are barbecue beef or lamb. But in most Han Chinese homes, aside from quick-fried, stewed in soy sauce, or simmered, the most popular way of preparing beef and lamb is none other than "rinsing" in a boiling hotpot.

All regions have dried or pickled vegetables, beans, eggs or meats. These foods intended for prolonged storage, are becoming less of a real course during meals as living conditions improve, but more of a tasty appetizing treat.

There are many ways to prepare Bean curd. It can be braised in soy sauce, or cooked in mala (hot and mouth-mumbing) flavor. (Photo by Roy Dang, provided by Imaginechina)

Seasonal Delicacies

Dumplings are a kind of folk food with a long history, and are loved by the common people. Just as the old saying goes, "Nothing tastes better than dumplings." For a long time in the past, having a meal of dumplings sometimes meant improvement of life. As for the history of dumplings, we can use the term "age-old" to describe it. The earliest recorded history associated with dumplings is found from the Han Dynasty. In the 60's of the 20th century, a wooden bowl was excavated from a Tang Dynasty (618—907 AD) tomb in the Xinjiang Uyghur Autonomous Region. In the bowl were wholly preserved dumplings, thus are the oldest dumplings ever found as of today.

Since ancient times, there has been a whole set of customs associated with eating dumplings. Dumplings on the night of the Chinese New Year and the fifth day of the first month on the lunar calendar, as well as on the day of "high heat" (beginning of the solar term of the same name from mid to later part of every seventh lunar month) and the first day of winter (around the 22nd of the twelfth lunar month). The saying 'There's nothing more delicious than *jiaozi*' bears witness to people's enthusiasm for filled dumplings in that part of the world. Many years ago, the phrase 'to eat a meal of *jiaozi*' was the same as saying 'give your life a turn for the better.'

Since the Ming and Qing dynasties, the folk custom of eating dumplings on Chinese New Year was already very popular. Especially in the North, until this very day, wrapping and having dumplings on Spring Festival is an indispensable feast activity for every family. On Chinese New Year's Eve, the whole family sits in a circle, kneading the dough, mixing the filling, rolling the wrap, wrapping, pinching, and boil the dumplings, all the while having a good time. This meal of dumplings is different from all other dumpling feasts throughout the year. After the dumplings are made, people wait until the clock strikes midnight before eating them. This makes the dumplings the first meal of the year. The Chinese word for dumplings, *jiaozi*, has the meaning of bidding farewell to the past and welcoming the new.

Dumplings are a special kind of holiday food, and also a common home-cooked dish. It can be prepared by boiling, steaming and frying. But the stuffing is what differentiates the type of dumpling. When having dumplings at home, pork stuffing is the most typical, and can be mixed with any vegetable. Pork is minced, mixed with sesame oil, scallion, ginger, and soy sauce for marinade. Right before wrapping the stuffing, mix already-made minced or ground vegetables into the stuffing and then add salt. Some prefer minced lamb or beef, but the most refined stuffing would be the *sanxian*, or literally the "three fresh meats," which is the combination of

Festivals and Food

Under the Qing Dynasty, the feudal dynasty nearest to modern times, the most important of the festivals celebrated throughout the country were those listed below. The food traditions attached to them have remained influential into the present.

Lunar New Year
jiaozi, sweet rice-flour dumplings

Beginning of Spring
spring pancakes, spring cakes

Dragon Boat Festival
zongzi

Mid-Autumn Festival
moon-cakes, gourds

Double Ninth
chrysanthemum wine, flower cakes

Winter Solstice
wonton

sea cucumber, shrimp and pork. To make a good dumpling, the wrapping is almost half the job. The amount of flour and water must be mixed in just the right portions, and the dough must be kneaded enough to achieve the best elasticity and best tenderness in the wrap. The wrap should be easily pressed together and must not be punctured easily. The cooked dumpling should be tender, juicy, and slippery with a tantalizing aroma after taking a bite.

Making dumplings is a time-consuming job. Therefore in recent years, already made dumpling wraps and stuffing are available on the market. Supermarkets supply cook-and-serve frozen dumplings in all kinds of flavors. Making and having dumplings has never been so easy.

In the southern part of China, the first meal of the New Year is usually not dumplings. Instead it would be stuffed glutinous rice balls, glutinous rice flour cake or noodles. China's numerous

Having dumplings for the New Year is a tradition in many families of the north. (Photographed in 1962, provided by Xinhua News Agency photo department)

Kodak
伊利冰淇淋

A New Year's painting Celebrating the Lantern Festival portrays the festive scenes of igniting firecrackers and making *yuanxiao* (stuffed glutinous rice-balls) among Chinese people. (From the collection of Shucun Wang)

minority nationalities also have the tradition of celebrating Chinese New Year, but with their own unique set of festival foods. The Hui people eat noodles and simmered meat on the first day of the first lunar month. The Yi people have "Tuo Tuo Meat" and drink "Zhuan Zhuan Wine." The Zhuang nationality likes to have a large sticky rice cake that weighs more than 2.5 kilograms (5.5 pounds). The Mongolian minority people gather around the fire to have boiled dumplings, but must leave lots of leftover wine and meat, only then will the coming year be full of prosperity.

The festive atmosphere of the Chinese New Year will last for a half month, until the 15th day of the first lunar month when it is the Lantern Festival, another important holiday of the Chinese. On the night of the Lantern Festival will be the first full moon of the year. Festive decor and bright lights adorn the major streets and narrow alleyways. The people while guessing fun riddles placed inside beautiful lanterns, enjoy *yuanxiao* (glutinous stuffed rice balls in soup), the southerners call them *tangyuan*. The main ingredient of *yuanxiao* is sticky rice; being highly glutinous, one must chew it

< Taking joy in watching lanterns on the Lantern Festival, 15th of the first lunar month. (Photo by Ren Jiang, provided by Imagechina)

thoroughly and cannot eat too much all at once.

The types of *yuanxiao* and ways of eating them are numerous. In the north, mixed stuffing made from osmanthus, rose, bean paste or sesame seeds would be rolled around in dry flour, until the stuffing is covered entirely in flour, taking the shape of a round ball, thus the *yuanxiao* is made. Salty *yuanxiao* is however very rare. In the south, *tangyuan* is made by putting stuffing directly into already formed dough, with sweet, salty, meat and vegetarian flavors available.

On the firth day of the fifth lunar month, the Dragon Boat Festival, *zongzi*, or glutinous rice cake wrapped in reed leaf is the festivity food. This custom is prevalent in all parts of China. The Dragon Boat Festival has had more than 2000 years of history in China. By tradition, people place portraits of Zhongkui, the demon-chaser, on the doors and walls of their homes to ward off evil, and hang up mugwort leaves. Grown-ups enjoy yellow wine while children play with "fragrance bags," used as protection charms. However, *zongzi* are had in both the south and the north, just with different flavors and shapes. The northern Chinese like to use jujubes, bean paste, preserved fruits and other sweet things as filling, coated with a thick layer of sticky rice and using reed leaves to wrap it into a triangular shape. In the south, there are also square and flat *zongzi* available. The fillings in the south are more abundant, with eggs and meats. There are sweet and salty flavors, each with its own satisfying tastes.

A farmer with pig legs on the shoulder going home to celebrate a festival. (Photo by Jian Zhu, provided by image library of Hong Kong *China Tourism*)

Second to the Spring Festival in importance and grandness is the Mid-autumn Festival, when people get together to have the moon cake. Like eating *zongzi* on the Dragon Boat Festival, or having *tangyuan* on the Lantern Festival, having moon cakes at Mid-autumn is a global Chinese tradition. Because of its round shape like that of the moon, the moon cake symbolizes union. Every Mid-autumn, when the bright round moon is hanging above, and all homes are united, people enjoy moon cakes while observing the moon and enjoying life. Moon cakes and *zongzi* are both considered desserts and not main meals. Even as such, moon cakes come in many flavors. There are over a dozen kinds including the five nuts, lotus seed paste, egg yolk, bean paste, crystal sugar, sesame seed, ham and more. They either taste sweet, salty, salty and sweet, numb-hot, and more. Traditional Beijing-style moon cakes are made similar to the sesame seed cake, where the outer crust is crispy and delicious. The Suzhou-style moon cake also has a crispy outer crust, and consists of many thin layers, soft and light, pleasing to the tongue. The Cantonese-style moon cake's skin is similar to Western cakes, but its inner filling is the most famous. Not only this, modern packaging of moon cakes are becoming more and more refined and eye-catching.

Aside from "the big four" on the Chinese calendar of festivities, which hold the most importance, there also exist certain seasonal folk customs with very interesting features. For example, in some areas, on the second day of the second lunar month, people eat Dragon Whiskers Noodles. On Pure and Bright Festival (Tomb-Sweeping Day), lighting fire to cook food is a taboo and people should eat cold dishes. On Festival of the Dead Spirits, dough-made man and sheep figures are offered to ancestors. On Double Ninth Festival, huagao cakes are had to bless the elderly with long life. There are still many more

Zongzi (glutinous rice cake) with meat stuffing (Photo by SCMP, provided by Imagechina)

A huge billboard ad for moon cakes on the street during mid-autumn. (Photo by Xi Yang, provided by Imagechina)

interesting holiday diet customs.

When it is once more near the end of the year, on the eighth day of the twelfth lunar month, people all around China have the tradition of eating laba porridge (rice porridge with beans, nuts and dried fruit), with slight variations in the making. Northerners prefer various kinds of rice and beans; southerners will add lotus roots, lotus seeds, water chestnuts and more. It matters not whether it is northern or southern style, red dates and chestnuts are a must. The Chinese word for jujube, zao, has the same pronunciation as the word meaning "early." The word "chestnut" in Chinese, li, also sounds just like words for "power" or "strength." With zao and li together, they imply the meaning of "putting in work and effort early in order to ensure a good harvest." As living standards improve, porridge ingredients are becoming increasingly abundant and varied, including peach kernel, almond, sunflower seed, peanut, pine nut, brown sugar, grape and so on. Laba porridge is now even more delicious with more refinement and higher nutritional value. Top grade laba porridge has therapeutic effects ranging from benefiting the spleen, stimulating appetite, replenishing qi (chi), cleansing blood, fighting cold weather and more. It is a characteristic tonic food for the winter season.

Interesting Food Customs

China is a country with great ethnic diversity. Subject to influences from geographical environments, climate, natural resources, as well as religious faiths, social and historical elements, every minority nationality formed their own unique customs in food and drinks. For example, those minority nationalities that rely on livestock are accustomed to eating beef and lamb, as well as the animals' milk and related dairy products, such as milk tea. Whereas for agricultural minority nationalities, those in the south rely on grains as principal food, the northern groups mostly eat pasta and mixed rough grains. Those living in frigid climates enjoy having garlic, while those living in the humid regions prefer hot foods. The Hui and Uyghur nationalities believe in Islam. To them, pork,

A traditional wedding of the Miao nationality in Hainan Province. (Photo by Yue Gu, provided by Imagechina)

meat of vicious animals and dead animals are forbidden. The Zang nationality, or the Tibetans, is barred by their faith from eating fish. If one does not know of these customs and prohibitions, feasting with minority nationalities can result in awkward situations.

Many people have heard of such a story, "A traveler on a horse trudges through the endless plains of the Mongolian grassland while carrying a whole leg of lamb on his back. At sunset, he sees a Mongolian tent, so he dislodges from the horse to stay the night there. The tent's master invites him in and put his leg of lamb aside. Then the tent master goes to his own sheep pen to bring out one of his own lamb for treating the guest. After the meal, the man and the tent master's family sleep in the same tent. The second day, the tent master bids the traveler farewell and gives him a new leg of lamb." The traveler has ventured far in the grasslands, and yet every time he leaves from a family, he would always be carrying a leg of lamb. But it is no longer same leg of lamb as it has been replaced with fresh legs countless times.

This story, by reasoning, should be true. The Mongolian nationality's warm hospitality is the most famous among all the minorities. Plus the lamb is the Mongolian nationality's main food for treating guests. According to local customs, regardless of loosely related kin, close neighbors, frequent guests or first timers, all guests will be treated to freshly slaughtered and prepared lamb. The sheep is first presented to the guest, and only after the guest's acknowledgement would the sheep be slaughtered. This is called "asking the guest, slaughter the sheep," and is intended to show respect for guests. Out of all the ways of eating lamb, "hand-served lamb" is the most characteristic of the Mongolian nationality's traditions.

"Hand-served lamb" speaks of water-boiled lamb without any seasoning. After being cooked, large chunks of succulent lamb, thick with juices and oils, give off steamy appetizing aroma. The local Mongolians like to eat it with one hand clutching onto a large piece of meat, while cutting it with their Mongolian daggers in the other hand. If highly honored guests arrive, a feast with a whole lamb must be prepared. This is also referred to as "yang bei zi," where a whole lamb is boiled in a pot. For the locals, preparation time needs to be only 30 minutes. When the knife cuts into the meat, blood would still seep out. If the feast were prepared for Han Chinese, it would usually have to be cooked for an extra ten to fifteen minutes. Meat cannot go without wine, as the Mongolians are big drinkers regardless of gender. At a feast, the host pours three bowls full of wine, while holding onto a white piece of Hada (long pieces of white silk symbolizing purity, loyalty and respect), proposes a toast to the guests while singing the toast song loudly to show sincerity. According to Mongolian customs, the guest should dip the tip of his or her middle finger into the wine, and flick once towards the heavens and once down at the earth to pay respect. Then it is "bottoms up" time. Trying too hard to decline the drinks would be viewed as a lack of sincerity.

A Tibetan woman making food. (Photo by Bing Shi, provided by image library of Hong Kong *China Tourism*)

Xizang, or Tibet, with its unique highland scenery and local customs attracts more and more tourists from China and abroad. The Tibetan nationality's food and drink habits are also a delightful attraction for the tourists. Those who have been to Tibet would surely have tasted its buttered tea. The Tibetan nationality uses buttered tea to treat guests. The guest must first have three bowls of the tea. If one wishes to have no

A banquet scene during a funerary ceremony of the Baiku Yao nationality. (Photo by Liyu Xu, provided by image library of Hong Kong *China Tourism*)

Young girls of Mosuo nationality must stand on fat pork as part of their coming-of-age ritual. (Photo by Xuezhi Li, provided by image library of Hong Kong *China Tourism*)

more, then one should pour the tea dregs on the ground. Otherwise the host would keep persuading the guest to have more. Principal foods of the Tibetans include vegetable noodle, buttered tea, beef, mutton and dairy products. A Tibetan family's wealth depend s on its amount of grain storage, not meat or milk, which every family has loads of. Tibetans usually would not eat horses, donkeys and other animals belonging to the Perissodactyla order, meaning odd-toed. Fish, chicken, duck, goose and other poultry are also not on their menus. Instead, they prefer to eat meat of Artiodactylas, the order of hoofed animals with even numbers of toes, including pork, beef and lamb, especially dried beef. In the Tibetan highlands, food will not mold or decompose easily. Dried beef jerky, being able to preserve freshness, are very common in the Tibetan region. Every autumn, Tibetans cut fresh beef into strips and string them together, adding salt and powders of Chinese prickly ash, hot pepper and ginger, and hang it to dry at cool, vented areas. Its texture is crispy and the taste is of long-lasting fragrance, with touches of sourness.

China's southwest region is one of its main concentrations of minority nationalities. There are many minority groups here where dietary habits are also multifarious and vivid. Humid climate here makes sour and hot tasting foods, as well as dried, smoked, cured and other preserved foods the preference of the people.

Women of the Wa nationality pounding rice. (Photo by Guozhong Liao, provided by image library of Hong Kong *China Tourism*)

Spread out through provinces like Yunnan, Guangxi, Hunan, Jiangxi, Guangdong and Hainan, the Yao nationality often add corn, millet, yam, cassava, taro root and kidney beans into rice porridge or rice. Since they often plough lands in the mountains, foods they make must be easy for carriage and storage. Therefore, the sticky rice cake and bamboo tube rice, both principal and supplementary foods at the same time, are their favorites. When working in the fields, Yao people all like to drink alcohol. In most Yao families, there are rice, corn and yam wines. Having alcohol two to three times a day is very common for the Yao people.

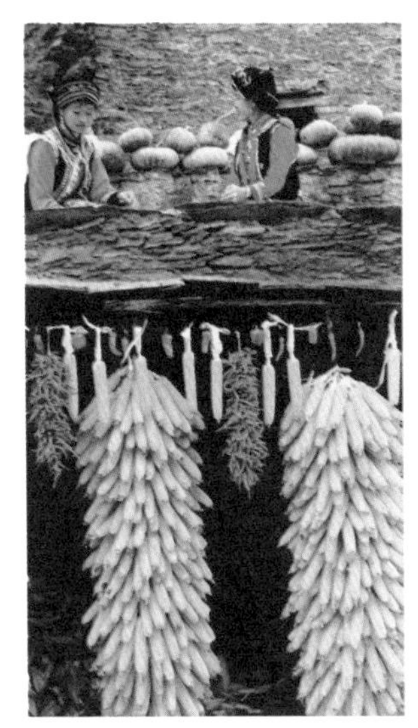
The upper floors of houses of the Wa nationality are used as storage space for foodstuff and other sundries. (Photo by Zhizhao Ye, provided by image library of Hong Kong *China Tourism*)

The Miao nationality, which populates much of the border regions of Guizhou, Hunan, Hubei, Sichuan, Guangxi and other provinces, commonly prefer sour-tasting dish, with Sour Soup in every family. The making of Sour Soup is by mixing rice soup and bean curd into an earthenware pot for three to five days until it ferments. It is used to cook meats, fish, and other vegetables. Preservation of food is commonly done by way of curing, for making sour vegetables, chicken, duck, fish and other meats. Almost every family has a "sour pot" for storage of cured foods. The Miao nationality has a long history of winemaking. There is a very comprehensive set of procedures from making

Men of the Wa nationality having lunch on the grain field. (Photo by Xianming Zeng, provided by image library of Hong Kong *China Tourism*)

wine yeast, to fermentation, distillation, and storage in a cellar.

The Dong nationality of Guizhou also loves sour foods. There are pickled Chinese cabbage, pickled bamboo shoots, cured pork and fish in every family. There's a Dong folk song that goes like this, "The man shall not be lazy; the lady shall not be over-indulged in having fun. Grow and cook good sticky rice and make excellent marinade fish. The mountains are full of treasures for diligent people, and every family has a full pot of sour foods." Furthermore, the Dong people's preserved duck and meat paste, fish and pickled ginger are all very well known. Interestingly for preserved fish, it must be sealed for storage underground for three years, sometimes even up to seven or eight years, before unsealed for use.

The Bai nationality is the most focused on foods for festivities among all the minority nationality groups. Almost every holiday has several holiday foods to go with. Ding-ding Candy is for Spring Festival. Steamed Cakes and Bean-starch Noodles are for the Third

Month Holiday. Pure and Bright Festival (tomb-sweeping day) has cold jambalaya with dressing and fried crispy pork. *Zongzi* (sticky rice cake wrapped in reed leaf) and realgar wine are for the Dragon Boat Festival. Various types of sweets and candy are available for the Torch Festival. For Mid-autumn Festival, it is the "White Cakes" and "Drunken Cakes." Lamb is special food for the Double Ninth Festival. There still much more holiday food traditions of the Bai people. It is truly a bright and colorful life.

The Zhuang nationality has the largest population among the minorities. They mostly inhabit the province of Guangxi, with also small populations in Yunnan, Guangdong, Guizhou and Hunan. The Zhuang-inhabited regions are teeming with rice and corn, which is naturally their principal food. The Zhuang people are not forbidden to eat any poultry or livestock meats; some areas even prefer dog meat. They often cook chicken, duck, fish and vegetables to about a medium degree, then just quickly stir-fried it before use to seal in the fresh taste. Rice wine is the main hospitality drink of the Zhuang people. Mixing the wine with animal innards, and we have Chicken Gallbladder Wine, Chicken Innards Wine, Pork Liver Wine and so on. When having Chicken Innards Wine and Pork Liver Wine, the drink is to be swallowed at once first, with the chicken innards or pork liver left in the mouth to be chewed slowly. These wines can be both a drink and a food dish.

The three provinces of China's northeast region are also inhabited by a few minority nationalities. The Korean nationality is a representative here; their food pays attention to freshness, fragrance, crispiness and tenderness. Most Korean foods have a piquant flavor. The ingredients in Korean dishes are usually the most tender of fresh meats cuts. It is usually raw with marinade, preserved, or boiled in soups. Raw and marinade beef threads, raw tripe, and raw fish slices are all traditional fare of the Korean nationality. Preserved vegetables of the Koreans are famed in history. Its ingredients are common vegetables, including Chinese cabbage, turnip, hot pepper, ginger and so on; salt is added before curing. The taste is refreshing with a stinging pungency, but with all five flavors present, being fragrant, sweet, sour, hot and salty at the same time. Korean pickled foods make fine complements to Chinese folk foods.

The Hezhe minority nationality that lives in the Sanjiang Plains area of Heilongjiang Province is the only group in northern China to rely mostly on hunting and the use of dogsleds for living. Their diet is rather old-fashioned, keeping tradition of eating raw foods until this day. Its most distinguished dish is the "Kill Raw Fish," where raw fish is mixed with vegetables rinsed with boiled water, including potato threads, mung bean sprouts and leek, added with chili oil, vinegar, salt and soy sauce. It gives off a fragrant, fresh taste and tender texture. In the Daxing'an Mountains live the Elunchun and Ewenke minority nationalities. Being amidst their

A Uyghur young man is grilling mutton kabobs. (Photo by Frederic J. Brown, provided by Imagechina)

"natural zoo" environment, they have kept the primal dietary tradition of "eating meat and drinking milk." They can often have deer milk and meat, roe deer feast, snow hare meat, pheasants and other wild animals. But these wild foods are already extremely rare delicacies in the inland part of China.

The Hui people who believe in Islam can be found throughout the country. They mix in with the Han Chinese, but adamantly keep to their unique dietary habits no matter where they go. Rice and pasta are their principal foods, with a preference for dough cakes, flapjacks, stuffed buns, dumplings, soup noodles, and noodles mixed with sauce and toppings. Compared with the Hans, the Hui nationality's biggest tabooed food is the pork, with dog, horse, donkey and scale-less fish also among the list of forbidden foods. They will not eat any meat of animals that were not slaughtered but died due to some other cause. Alcohol is also strictly forbidden. Since the taboos are strictly enforced, in towns and cities, the Hui people have their own qingzhen restaurants, so they would not have to dine with other non-Muslim people. Therefore, Hui qingzhen food stands unique among the numerous minority nationality food styles, and has produced many qingzhen dishes such as Triple Quick-fry, Steamed Lamb, Lamb Simmered in Yellow Sauce, and Lamb Tendons, which are all famous fares. Names such as Donglaishun, Hongbinlou and Kaorouji are all very famous qingzhen restaurants in China and even on the international scene. It is safe to say that the development of Hui qingzhen food has made great contributions to Chinese diet and culinary arts as a whole.

The Rituals of the Table

China has always been credited with being a "State of Etiquettes." According to historical documents, as early as 2,600 years ago, this nation has already established a very comprehensive set of dining etiquettes. The ancient Chinese, when treating guests to a banquet, must prepare adequate seating for the quests. For a

Moral Virtue and *Five Things to Ponder when Eating*
Although the Chinese do not say grace before meals, they do have a tradition of using meal times to examine and reflect on personal thoughts, words and deeds. Almost as soon as a child starts to speak, adults teach him or her to recite the words of the old poem 'For each grain of food on the dish, who knows how much toil lies behind it?' And the 'Five Things to Ponder when Eating' written by Huang Tingjian in the Northern Song period is still used as a personal yardstick by some people. The 'five things' are these. 1) When eating or drinking, be aware that these things are precious and not easily obtained; 2) Consider whether your own behaviour has been good enough to justify the food and drink you have, and if you have any shortcomings you should feel ashamed – you cannot be greedy without a thought; 3) for the better improvement of your body and mind, when eating and drinking you should avoid greed, anger and giddiness; 4) you should know the nutritious properties of the various grains and vegetables, and use this understanding to guide your practice; 5) make sure that under all circumstances you have high ambitions and ideals –always have something to contribute which is on a par with what you consume.

large number of guests, the elderly or people of high status get separate seating. When occasionally sitting with others, they are also to sit at the head seats. For seats that are arranged with north-south orientation, the seat to the west is the head seat. East-west seats have the southern seat as the head seat. People sitting on the same seat must have comparable status, or it is considered disrespectful. Before guests can be seated, the seats and tables must be properly oriented and aligned; otherwise it must be adjusted before guests can sit down, or it would be observed as infelicity. Regardless of host or guest, one must hold a peaceful expression in the face, and lift one's long sleeves, with both hands, to about one chi (Chinese measurement unit) from the ground when taking a seat. After being seated, one's upper body garments must never be lifted and the feet are not to move about too freely.

A famous 19th century Russian writer, Anton Chekhov, once invited a Chinese man to have a drink in a bar. Chekhov said, "Before drinking from his cup,

Weddings in Chinese cities usually are held in combination of Chinese and Western styles. The groom and bride will make toasts to the guests one by one. (Photo by Liqun Liu, provided by Imagechina)

Rich sacrificial rituals among local people in Shaanxi Province. (Photo by Yankang Yang, provided by image library of Hong Kong *China Tourism*)

he held it with his hands and presented to me and the tavern owner and bartenders, saying 'qing (please).' This is the custom of China. They are not like us to drink it down in one gulp, but prefer to sip on it little by little. With every sip, he ate some food. Afterwards he handed me some Chinese coins to show gratitude. This is a rather interestingly polite nationality..." This was the opinion of a Chinese person given by a foreigner two centuries ago. Chinese traditional banquet formalities were numerous and tedious; the more grand the occasion, the more detailed the formalities were.

These rules of conduct have passed on to the present day, though it has changed somewhat in format. At rather formal banquets and feasts, a round of courtliness cannot be avoided before all participants take their seats; the seating order and arrangement are also disciplined. Usually, the highly respected, the elderly, the host or the host's most treasured guest sit on the north side of the table facing south or directly facing the entrance door to the room. The order of seating starts normally with the elderly; the wedded has priority over those that are still single; the non-familiar guests take seats before the close friends and acquaintances do. But depending on the occasion, the order of seating is somewhat different. On the birthday of a senior, the "head seat" belongs to that person. The sons and daughters, either natural or in-laws, sit at the east and west of the head seat. For the one-month-old celebration of a newborn baby, many regions of China like to leave the "head seat" for the child's maternal grandmother. As for wedding banquets, the "head seat" is reserved for the uncle (brother of the mother) of the bride.

The Chinese, when eating at home, do not encourage having alcohol for every meal. But on a banquet, alcohol drinks are a necessary item. After the guests take seats, the host must propose a toast to the guests while saying "drink first to show respect," then the host and guests all drink up. Regardless of host or guest, refilling must be to the tip of the cup. For those with low alcohol tolerance, it would be wise to declare beforehand in order to avoid awkward situations.

Respecting the senior and cherishing children are traditional Chinese virtues. A square dining table makes people feel strong human bonding. (Photo in 1960, provided by Xinhua News Agency photo department)

There is a set of etiquettes when serving plates of food. The dishes with bones are placed on the left side of the table; dishes with meat are on the right. Rice and pasta are on the left-hand side while soups, alcohol and beverages on the right. Roasted meats should be placed further away from people, with seasoning including vinegar, soy sauce, scallion and garlic near the diner. The order of serving is from cold to hot. The hot entrees should be served starting on the left of the seat across from the main guest.

When eating, one must obey the rules, commonly referred to as "diner's appearance." For example, one must not poke their chopsticks vertically into the center of the rice bowl. When finished with the meal, one should not say, "I'm done," but should rather say, "I've eaten well" or "I'm full." Avoid making sounds with the chopsticks tapping the bowl and so on. The Chinese from childhood are taught to "stand properly, sit properly and eat properly," and accept all kinds of "dining appearance" training. The trainings include how to select seats; how to deal courtliness to the three kinds of people with priorities; how to hold chopsticks; how to pick up food with chopsticks; when to talk cheerfully and humorously; and when to be taciturn. As a result, even dining for children is not a simple matter. They do not dare to leave unfinished grains of rice in the bowl, because of the fear of getting pockmarks on their faces when they grow up. When clipping food with chopsticks, one cannot pick up too much at once; even worse is to poke at foods too often or stir around inside a dish. When eating, one should not make noises with the mouth and cannot devour ravenously. When eating soup, one cannot slurp or be sloppy, and the list of etiquettes goes on and on. "Dining appearance" is a propriety that the Chinese, especially the senior generations, pay much attention to.

Modern day Chinese, particularly the youngsters, cannot help but feel these formalities too tedious and restrict freedom. But it also cannot be denied that it is because that "proprieties" exist, a banquet can proceed with order, and that hosts and guests are able to communicate their feelings and thoughts. At the same time, it

Knowing the Flavours
The Chinese use the term 'the five tastes' to refer to the sweet, sour, bitter, hot, and salty characteristics of food, and these are the five basic flavours. Healthy people are able to discriminate between these five flavours in food and drink. The classic text 'The Doctrine of the Mean' from the 5th century BC puts forward the idea that 'while everybody eats and drinks, few know the flavours'. So 'knowing the flavours' is seen as the food connoisseur's field and expertise. There are some stories about connoisseurs who were especially good at distinguishing between flavours. In the Pre-Qin period there was a chef called Yi Ya who could tell the difference between the flavours of waters taken from two different rivers; there was a blind musician called Shi Kuang, who while dining with his king discovered from the taste that old timber had been used to cook the rice, and when the king thereupon checked with his chef he was told that the axle of an old chariot had indeed been used as fuel; there was also a nobleman in the Jin period, Fu Lang, who was so discriminating that he could tell whether the chicken he was eating was free-range or had been kept in a coop.

keeps uncivilized acts to a minimum. Etiquette, after all, is the bridge to high culture.

Balancing the Five Flavours

If we say that the goal of eating and drinking is to improve health, then the number one important element in food would be nutrition, demonstrating a scientific practicality. The Chinese focus on color, fragrance, taste and form in food, looking for refinement in food vessels and elegance of dining environment, demonstrating an artistic spirit. Hence ever since long ago, the Chinese advocated the philosophy of "five tastes in harmony." The Chinese invented ways to adjust blended ingredients and spices for a wide variety of tastes. Revolving around the "five tastes," which are sourness, sweetness, bitterness, pungency and saltiness, dishes can evolve into more than 500 different flavors.

Of the "five tastes," saltiness is the principal taste. It is the most simplistic and most crucial. Salt is needed to heighten any texture in foods. Without it, any delicacy cannot emerge in its full glory. But from a health perspective, salt should not be taken in excessive quantities.

Sourness is also an indispensable taste in foods, especially in the northern part of China, where water supply is heavy in minerals and strong in base. So in order to induce better digestion of food, vinegar is often used in cooking; and it could also arouse appetite. Sour taste can also neutralize fishy odor and greasiness. At banquets with strong grease and heavy meat dishes, sour dishes are usually complemented and come in many varieties. Not only are the sour

Engraving from the Jin Dynasty, portraying the general process of production of salt by boiling in the Song Dynasty.

tastes of plums, fruits and vinegar different from one another, just the different type of vinegar are distinguished by its production areas, different ingredients and different techniques of making, thus causing quite drastic differences in taste. Usually, the northerners uphold mature vinegar made in Shanxi as orthodox, whilst the people in the Jiangsu-Zhejiang area appraise the Zhenjiang-made rice vinegar as authentic. The most typical of all places eating vinegar is the province of Shanxi. Many families there are skilled at making vinegar from crops and fruits. Their everyday meals are even more dependent on vinegar. A very interesting thing is that in the Chinese language, the word "vinegar" is used to represent the feelings of jealousy between men and women. Slang, such as "eat vinegar" and "vinegar jar," are universally understood in both the north and the south. It may have to do with the sour nature of vinegar itself.

Pungency is the most stimulating and complex of the "five tastes." Sometimes we use "pungent-hot" as one word. In actuality, pungency and hot has major differences. Hot is sense of taste, stimulating the tongue, throat and nasal cavity. Instead, pungency

The illustration *Buying Water* printed on matchboxes at the beginning of the 20th century. (Photo by Zhongmin Lu)

Vegetable stands and fish stalls painted on matchboxes at the beginning of the 20th century. (Photo provided by Zhongmin Lu)

is not just a sense of taste as it involves sense of smell as well. Pungency is mostly obtained from ginger, while hot and spicy usually denotes the use of hot pepper or black pepper. Since hot peppers were once a foreign product, there were no mentioning of "hot" in ancient Chinese cooking, instead was generalized as pungency. Ginger not only neutralizes rank taste and odor, but can also bring out the great taste of fish and meats. So ginger is a must-have when making fish and meats. There are also principles to using hot peppers. We should not merely seek for the degree of hotness, but should rather use saltiness and natural essence of food as fundamentals, so that the hot and spicy taste comes out multi-staged, full of great aroma and not too dry. In addition, garlic, scallion, ginger and other spices can also kill bacteria, so are great for cold dishes with dressing.

Bitterness is rarely used alone in cooking, but is a valuable asset. When making simmered or braised meats, adding tangerine or orange peel, clove, almond and other seasonings with a light bitter touch can rid the meats of unpleasant taste and smell, and awaken the tastiness of the food. Traditional Chinese medicinal theories believe that bitterness is helpful for the stomach and produces saliva. Some people really enjoy bitter taste in foods, such as in the Sichuan-style "Strange Taste" type of foods, which have the bitter elements.

Sweetness has the affect to cushion the effect of other basic tastes, whereas saltiness, sourness, pungency and bitterness are all too strong, they could be remedied by sweetness. When making dishes of other tastes, sugar can improve and embellish. However, using large amounts of sugar is not recommended, as too much sugar can be nauseous. Since many spices can produce a sweet flavor and they all taste quite different, much of the culinary world hails cane sugar as the orthodox sweetness.

What is not listed in the "five tastes" but still holds important status in the culinary world is the "fresh essence" factor. "Fresh essence" is the most tempting taste in food. Most foods all contain the "essence" but it is often dormant, so making soup is often the way to awaken the taste. Chicken, pork, beef, fish and ribs can all be used as soup stock. When the unpleasant tastes and smell are eliminated during the soup-making process, the essential flavor is fully exposed by adding just a touch of salt. Essence soup not only can be enjoyed directly, but can also be used to make other plain foods taste great. Such foods include shark's fin, sea cucumber, bird's nest, bean curd and gluten, which all must be cooked with essence soup to achieve its mouthwatering taste. Monosodium Glutamate (MSG) is manmade essence. Its synthetic nature makes for it impossible to compare to naturally made essence soup. So

This Qing Dynasty folk illustration portrays a vendor of Tofu Brain snack (soft bean curd in soup). The food is placed at the rear load, while a wooden basket with a wooden board are in front; on it are bowls, spoons and seasonings. (Provided by Shucun Wang)

skilled chefs often would not care to use it.

Chinese cuisines shine at mixing and blending of flavors. This is not only aided by superior culinary techniques that can mix natural flavoring, but also with the help of a whole line-up of seasoning. Aside from salt, vinegar, sugar and essence soups, which are representative seasoning, pastes, soy sauce, wine, and stinky tofu are all commonly used seasoning in Chinese cooking.

Paste made from the fermentation of beans was regarded highly in ancient China. Once it was the food for the upper class. When treating guests at banquets, bean-sauces must be served, since each kind of meat has its matching paste. Experienced eaters will know the kind of great food to be served just by seeing the type of paste. In time, pastes became important seasoning, from which a whole series of seasonings were developed, including soy sauce, bean-paste, black fermented beans and more. Pastes made from beans are very much a Chinese specialty sauce. It holds an important place in Chinese culinary history, or even the culinary timeline of the entire world.

Using wine for the blending of taste is also a great invention of Chinese cooking. Wine not only kills the rank stench of fish and meats, it can also produce a real appetizing aroma. When making stir-fries, adding a little cooking wine can bring out the delicious aroma of the food within the evaporating wine; the texture of the food is melt-in-your-mouth tenderness.

Aside from the Chinese, it is not known whether there are other people in this world who enjoy food with a stinking stench. Cheese from the Western world seems to border on foul smell, but compared to the Chinese's stinking tofu, the smell of cheese is nothing but child's play. Stinking tofu smells awful, but taking a bite makes all the difference; it becomes wonderfully delicious. Northern and southern China produces stinking tofu of different flavors, and stench. The northern kind is mainly used as sauce, while the southern type is a complete dish in itself. From ingredients to the making, it is all quite a refined process.

Picking out seasonal vegetables. (Photo by Huiming Shi, provided by Imagechina)

Chinese culinary techniques are an art of taste. A monotonous flavor gives a sense of imperfection to the taster. So the five tastes must be mixed and blended to mutually make good use of one another's strong points, leaving the taster with an everlasting and satisfying aftertaste. In actual cooking practices, the chef must be flexible in mixing the flavors to not only suit the diner's preferences and the seasonal characteristics, but to keep it healthy as well. Take salt for example, on a table full of courses, the first dish to be served will have normal amounts of salt added. However, salt is to be reduced gradually as we approach the last served dish. The soup that comes at the end is usually void of salt. Of course the diners do not notice the subtle differences and only feel the food is suited to their tastes. Cuisine styles of different taste categories still use similar ingredients, and cooking methods are all the usual stir-fry, fry, steam, boil and so on. The main difference lies in the blending of tastes. Taste blending is a very subtle and delicate art, the portions of different seasoning, order of application, and timing (before, during or after cooking) all must be just right. First or last, more or less, subtle differences but big on rules. Too early, too late, too much and too little, all would not do. When people say they like certain food, what they really mean is they like its taste.

There is no universal good taste, as every person has different preferences. Some people like natural flavors and juices, such as lightly simmered or steamed. Chicken must retain the original chicken flavor while ducks must taste like ducks. There are also people who like "strange taste" chicken and duck. Some prefer a thick, strong taste while others like it light and mild.

Modern Chinese, especially urban people, are becoming more and more mild-taste-oriented. Yue style, or Cantonese food, which emphasizes original taste and natural, tender texture, seems to fit this growing trend. In making Cantonese cuisines, usually no strong vinegar and soy sauce are used; only very small amounts of oil, salt and sugar are added. The dishes rely on the natural fresh essence that the food carries, so not crossing the fine line of the perfect level of seasoning is the key. This kind of taste preference by the urban Chinese most probably has to do with improvements in living standards. In the past when food supply was scarce, and freshness preservation technology for food was limited, using strong seasoning to make up for the lack of fresh flavors in food was the only way. Nowadays, "thick soups, heavy taste and ample oils" are a thing of the past, and no longer the standards of a good tasting dish.

Moreover, due to differences in regional climates and living customs, taste preferences differ greatly. The Chinese also have the tradition and custom to mix and blend tastes in line with the seasons. In spring, when all plants are budding and everything starting anew, food is the most susceptible to bacterial contamination. When making cold food with dressings, vinegar and crushed garlic can be added to fight the bacteria. Summer time speeds up the dehydration process, so people like to eat foods with strong base or with a slight bitter taste, such as the bitter gourd or leaf mustard. In the fall, high calorie foods and hot and spicy foods should be had more often. In winter, high calorie but heavy tasting foods are a good supplement; salt intake can be adequately increased to assist the digestion of meats.

Five tastes in harmony, with flavor as the top priority, bringing direct pleasure to the tongue. At the same time, it is a good health-protecting and body-regulating method. Chinese traditional medicine theories state that pungency can regulate bodily fluids, blood and qi (chi), and can be used to treat bone and muscle pain from coldness, kidney problems and so on. Sweetness can nourish, soothe, and improve emotional mood. Honey and red jujubes are also great tonic foods for those who have a weak and frail physique. Sour taste can cure diarrhea and produce saliva to stop thirst. Sour vinegar can prevent colds, while eggs boiled in vinegar can stop coughing. All these are folk cures with adequate modern medical recognition. Bitterness can release heat in the body, improves vision and detoxify the body. Five tastes in harmony is an important factor to great health and long life.

All in all, the so-called "five tastes in harmony" should include the following three tiers of meaning. Firstly, every dish must have its own unique flavor. But on the table, different courses should have different flavors to complement one another, so as to give a balanced feel overall. Secondly, adjusting the thickness of flavor is very important. Seasonings must all be able to perform, so that the resulting flavor is ever changing from dish to dish. Thirdly, when eating, one cannot be inclined to eat excessively only the foods with one particular taste and ignore the others. "Harmony" is the essence of Chinese philosophy, having multiple meanings such as "harmonious," "peace" and "unison." "Harmony" is also the highest transcendence of Chinese culinary arts. "Five tastes in harmony" is a reflection of the Chinese's pursuit of moderation, equilibrium, and balance, respecting the forces of nature.

Secrets of Delicacies

In ancient times, people who cook for a career were called

'Chewing Roots' Builds Character
In the past, Chinese intellectuals favoured building character through simple living. They thought that a hard material existence could train and temper the will and foster virtue, and at the same time could act as a spur to a man's progress. The Ming period essay *'On Vegetable Root'* by Hong Yingming took as its starting point the quotation 'If a man has chewed roots, there are a hundred achievements within his grasp.' The essay records the philosophy of life that can be derived from savouring a normal three-meal-a-day life, and was not only much loved by Chinese readers of the time, but also became a popular book in Japan and Korea and other parts of Asia.

"pao," now we call them chefs. Compared to the internationally recognized Chinese dishes, most of those who create these delicacies are obscurely anonymous. In Chinese history, Peng Zu and Yi Yin are quite famous chefs. The first-ever documented chef, Yi Yin (dates unclear), was also a prime minister of the Shang Dynasty. He was not only a man of erudition but also an accomplished political-military strategist. In addition, his superb culinary skills gained him much trust from the rulers of the time. Every time when holding rituals at ancestral temples, Yi Yin would explain to the Shang King in great detail the study of food. He would mention everything from cooking to the names and descriptions of all the delicacies in the world, classifying them into different schools. Taking

A kitchen range in ancient houses in Anhui Province. (Photo by Xuezhe Xu, provided by image library of Hong Kong *China Tourism*)

the opportunity, Yi Yin would incorporate many nation-governing principles into his teachings about food, so the common people recognized him as "god of the kitchen." In later periods, there were also great chefs, through excellent culinary skills, that acquired high ranks and generous salaries. But after all, it is a rare fortune. Most *pao* were still just servants to the nobility.

Among the common folks of China, chefs have always been viewed as a respected profession. They rely on their own excellent skills to live in society. Those who service the common people are the chefs in public restaurants. In the old days, they were called "*shi chu*," or "city chef." As the food service industry developed, the classification in the roles of chefs became more and more precise, which resulted in titles such as cooking chef, pastry chef and so forth. The passing on of culinary skills is no longer through the singular format of one teacher to one student. Instead, as an important make-up to modern professional education and training, cooking is now taught in specialized vocational schools, from which the graduate student can receive the "Professional Qualification Certificate of the People's Republic of China." The majors of this profession not only must learn culinary skills, but also partake in basic nutritional studies. The promotion in ranks of the chefs requires further professional testing.

Housewives who take charge of cooking were called "zhongkui" in ancient times. Though they were not considered chefs, their hands were skilled, and can often make eaters drool. Also in ancient China, studying cooking was mandatory for women before marriage. Even though many modern women now have taken steps outside the family and into the professional world, being able to make a fine meal is still the blessed gift and pride of the family.

Speaking of a group of people who have contributed the most to Chinese food culture, we cannot leave out the connoisseurs and summarizers of food, who were the literati and great men of ancient China. It was through their documentations that the skills and secrets of the chefs could be passed down through the ages.

A chef making dishes in a qingzhen (Hui Muslim) restaurant. (Photo by Bihui Li, provided by image library of Hong Kong China Tourism)

Aided by their superb art appreciation levels and cultured tastes, Chinese culinary techniques entered the realm of art. Sometimes, these great men personally participated in the making of dishes. Su Dongpo (1036 – 1101 AD), the great writer of the Song Dynasty, not only was a man fond of eating, his creation, the Dongpo Pork became a highly praised and popular dish. These epicures and the chefs of the past together brought about China's rich and consummate culinary arts. Qing Dynasty poet Yuan Mei (1716 – 1797 AD) wrote in his work, the Suiyuan Shidan, detailed accounts of 326 dishes from the 14th to 18th centuries; ranging from exotic meat dishes and superb seafood, incorporating delicacies from both the north and south. This work is a valuable historical documentation in the history of Chinese food culture. Using elegant words of commentary and abundant knowledgeable of cooking, he was honored as an epicure of the highest taste and scholarship. However, in tradition, wealthy families all hire in-house chefs to take care of daily meals to grand feasts. Banquets prepared by wealthy families were usually held in their homes, rather than at public restaurants. So to be able to find and employ a first-class chef, for a family, it is something to show off to others. Under such prevalent social atmosphere, chef's skills made progress in leaps and bounds

Chinese cooking attaches greatest importance to huohou, duration and degree of heating. (Photo by Min Qu, provided by Imagechina)

A chef's cooking and cutting skills include the mixing of supplementary ingredients, knife work, duration and heat control, and the specific cooking method. In daily life, basic ingredients used to make meals include four categories: vegetables, fish and meats, eggs of poultry, and seasonings. The so-called culinary arts are mainly just adequate mixing and cooking of the four categories of materials. Having Chinese food is different from having Western food. For example, when ordering a Western steak, the diner would be asked to choose from rare, medium or well-done. The chef prepares the food fully according to the diner's demand and adds no spices or seasoning. After the steak is served, the diner adds salt, pepper, lemon juice or ketchup totally depending on personal taste preference. Chinese food is simply ordered by the menu. How the dishes are prepared, deep or stir-fried, boiled or steamed, rare or well-cooked, hot peppers added, vinegar drenched, how much oil and how much salt, unless the customer makes a request, everything will be left up to the chef. The same kind of dish by the same chef will not have much difference in the general making, regardless of the diner, the dish will basically be the same taste.

Blending of supplementary ingredients is a Chinese chef's primary skill, a basis to making tasty food. Supplementary ingredients must be refined and thoughtfully added while keeping in mind the original properties (place of origin, age of growth, which

People of the Dong nationality in Guizhou refer to kitchen range as "fire pond." (Photo by Guanghui Xie, provided by image library of Hong Kong *China Tourism*)

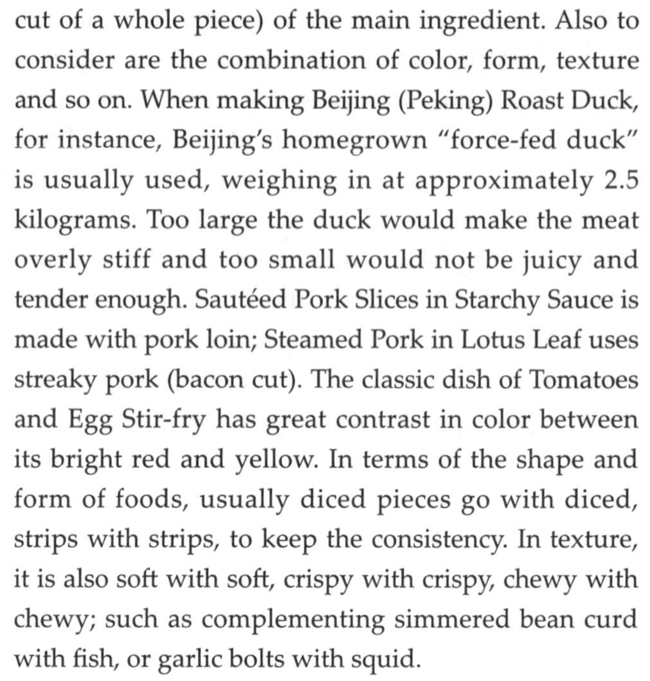

cut of a whole piece) of the main ingredient. Also to consider are the combination of color, form, texture and so on. When making Beijing (Peking) Roast Duck, for instance, Beijing's homegrown "force-fed duck" is usually used, weighing in at approximately 2.5 kilograms. Too large the duck would make the meat overly stiff and too small would not be juicy and tender enough. Sautéed Pork Slices in Starchy Sauce is made with pork loin; Steamed Pork in Lotus Leaf uses streaky pork (bacon cut). The classic dish of Tomatoes and Egg Stir-fry has great contrast in color between its bright red and yellow. In terms of the shape and form of foods, usually diced pieces go with diced, strips with strips, to keep the consistency. In texture, it is also soft with soft, crispy with crispy, chewy with chewy; such as complementing simmered bean curd with fish, or garlic bolts with squid.

Sometimes, special treatment must be given to the ingredient according to the style of cooking. Such as Hangzhou's famous West Lake Fish with Vinegar, which uses actual grass carp from local freshwater lakes. Though the fish is tasty, its meat is loose and with a slight aftertaste of soil. Therefore, before it is cooked, it must be placed in a specially made bamboo basket to be kept alive without feeding, so its meat will become tender and full of fresh flavor when prepared for food.

Without a doubt, the selection of ingredients for any dish sets benefits to health as primary consideration. Turnip has the power to relieve excessive internal body heat, so it is very fitting to be made into dishes with lamb, which causes high body heat. Spinach and tomatoes contain relatively more acidic substance. If it were coupled with calcium-rich bean curd, forming calcium salt (calcium hypochlorite), then it would be unfavorable for digestion and absorption.

The strength of heat and cooking time is what the Chinese call "huo hou," literally meaning, "fire degree." It is the most important link in Chinese cooking, and also the most difficult to master. Deep-fry or stir-fry requires strong heat, or the food will come out sloppy. When boiling food with water, moderate fire should be used. If heated under strong fire for an extended time, the food will be dried and shriveled. When frying, be careful not to cook for too long, or the food will be burnt and its taste altered. Making fish demands the most control in heating. The best-made fish should be pure white in color, its meat tender but not loose. For some foods, the more you boil it, the more tender and plump it becomes, such as eggs and kidneys. While other foods will immediately turn stiff if you overcook it for one extra minute, these include fresh fish, clams and other salt and freshwater foods. *Huo hou* changes all the time, without many years of on-hand practice and experience, cooking to the perfect degree can be difficult. Therefore, controlling cooking heat is a very important criterion of competition between

Chefs making dishes in a restaurant kitchen. (Photo by Qinghua Tang, provided by image library of Hong Kong *China Tourism*)

Chinese chefs. Whether one becomes a famous chef all depends on conquering this one barrier. Experienced chefs can freely make their food tender or crispy by will; can control the taste to be sweet, sour, salty, hot, mild or any combination there of. Take Stir-fry Pork Liver for example, the oil must be heated until boiling. When white smoke starts to fume, the liver goes into the wok. Immediately the liver is quickly stirred around, followed by adding starch. After just being tossed around for a bit, the liver is done. If the fire is not sufficient and the heat not strong enough, the liver will turn from tender to tough. Vice versa, if the fire is too strong and the wok is too hot, the liver will start to pop when dumped into the oil. The texture will also be affected.

Huo hou, when looking at it literally, means the degree and duration of the burning of fuel. But in cooking, it is not this simple. Ingredients, cooking utensils, as well as the material's heat conductivity are all concerned with huo hou. Modern cooking use natural gas or coal gas stove; ancient people used firewood, which has many specifications. Different types of firewood produce different tastes and have different effects on food. Using mulberry wood to roast duck or other meats will tenderize the meat faster and can detoxify. Rice spike can be used to cook rice, it is said that it can soothe people's minds. Using wheat spikes to cook can quench thirst and moisten the throat, and is beneficial to urination. Rice cooked with pinewood can strengthen muscles and bones, but pinewood is not suited for making tea, as tea requires coal fire. Using cogangrass can improve vision and detoxify the body. If making tonic medicine, reed or bamboo is the best choice for making fire. All such customs are still more or less preserved in rural areas, but are no longer feasible in the cities.

However, modern people have more choices for cooking utensils. Stir-frying needs to have concentrated fire and heat, so a round-bottomed wok is perfect. Deep-frying requires evenness of heat; a flat iron pan should do the trick. Some simmered foods needs to be cooked on low heat, such as simmered hen, turnip soup

A *qingzhen* (Hui Muslim) chef making hand-pulled noodles. (Photo by Guanghui Xie, provided by image library of Hong Kong *China Tourism*)

with ribs, or tremella and lotus seed soup, for which the Chinese usually use earthenware pot, or casserole.

Knife work refers to the ways a chef handles cutting of ingredients. The differences between Eastern and Western culinary skills are apparent when it comes to cutting. The Chinese meticulously cuts everything to desired shapes and sizes before the food is cooked; in the West, food is mostly cut further right before eating. It is obvious that Chinese chefs pay more attention to knife work, as all skills on the chopping board are most worthy of pride. For the ways of cutting, just the techniques with names amount to more than one hundred. Straight cut, leveled slicing, tilted cut, and retained cut (cutting but still keeping the ingredient partially intact), all are not easy to execute. Take Stir-fry Kidney Bloom for example, a very common dish that can be prepared with more than a dozen cutting techniques. The "kidney bloom" that is produced can look like an ear of wheat, a lychee fruit, the Chinese character of "*shou* (long life)," a comb, an orchid, a straw rain cape and more. With different ingredients, different cuts are executed. As the saying goes, "perpendicular cut for beef and cut chicken in the

direction of the meat fibers." Beef fibers are rather thick, so cutting perpendicularly to the muscle fiber will make it easier to cook. As for chicken, its muscle strands are rather thin, so the cut must be along its direction. This will preserve its tender and smooth texture, otherwise the meat will not be able to withstand any stirring in the pot, and will crumble to pieces.

Westerners cook simply by frying, boiling, baking or roasting on fire. In comparison, Chinese cooking uses abundant techniques, such as stir-fry, quick-fry, deep-fry, braise, simmer, boil, steam and so on, with no less than twenty ways. Every technique has its corresponding famous dishes. The most commonly used technique is still *chao*, or stir-fry. It is difficult for Westerners to comprehend, so the term "stir-fry" designates frying while stirring for constant motion. After all, westerners do not own "woks" which are specially made for stir-frying.

Most Chinese home cooking does not demand much from the wok, the main cooking utensil. Most often, a wok performs multiple functions, making dishes or boiling soups. However, for a professional chef, special stir-fry woks must be used. The wok has a handle for easy carrying and constant tossing of food. Tossing food with the wok can assure an even mix of ingredients, making the food just at the right tenderness and evenness. Tossing is a very complex skill. Colorful bits of food slices, strips and dices are gracefully tossed up through the air in an arc, and caught as they fall down in its original order back into the wok. This set of "stir-frying" skills is not executable with a flat pan, as it is another consummate skill of the Chinese chefs.

Stir-frying becoming the dominant technique of Chinese cooking was no accident. It is closely connected to China's general supply situation. When roasting, baking, frying, deep-frying, simmering, or steaming meat dishes, the size and quantity of meat usually cannot be compromised. But stir-fry gives much leeway in this area. When just a couple of tenths of kilogram of meat is minced, bits of turnip, Chinese cabbage, cucumber, peas, rice noodles, bean curd and the

< Mongolian women steaming buns of mutton stuffing in a Mongolian tent. (Photo by Xiuquan Chen, provided by image library of Hong Kong China Tourism)

like can be mixed with it to produce dishes that are inexpensive in ingredient cost, and still taste and look great.

Before the Han Dynasty, the Chinese had not developed the technique of "chao" yet. The main forms for food at the time were stews and soups, fire-roasted, boiled and fried. But as soon as stir-fry was invented, it dominated cooking due to mass acceptance. The word "chao" even became the general term for all forms of cooking performed with a wok regardless if it is deep-frying, frying, boiling, or steaming.

Color, aroma and taste are the joint standards that the Chinese use to evaluate the quality of foods. Only when the dish excels in all three areas can it be considered a well-made dish. Stir-frying can achieve this purpose the easiest. Stir-fry is not constrained by the quantity and type of ingredients; anything can be mixed and stir-fried in the same dish. This creates for a good basis of "color." Stir-fries are usually rich in oils, with finely cut ingredients cooked under high heat; the taste seeps into the food quite easily, emanating wonderful aroma. Especially when ground garlic or scallion sections are cooked as flavoring, the aroma is overwhelmingly strong but satisfying. Stir-frying also inadvertently aligns with nutritionists' teachings. Since stir-fries are made usually very quickly, it prevents the loss of nutritional content from food.

Opening our eyes to the great wide world, it can be seen that countries and regions that place strong emphasis on diet and have great food are all countries that have, at some point in history, reached great heights in cultural development, and have possessed great social-economic prowess. Only under such conditions can a country have enough time and money to pursue the fun and enjoyment of food and develop techniques of cooking. Refined culinary techniques allow for Chinese food to have unique tastes and charm. But this art of cooking is now being tested and challenged by our times. As the food processing industry becomes mechanized and automated, the very survival of the basic skills of Chinese culinary arts is being threatened. All heat-and-serve and frozen foods in supermarkets have gained solid market share, and more intelligent electric cookware is introduced. Much of home cooking can now be taken care of by automatic and procedural means. It seems that Chinese culinary skills are becoming somewhat obsolete, especially when facing those foods that are produced in large quantities. Nevertheless, the living habits of the Chinese, specifically their pursuit of color, aroma and taste in foods, have destined the Chinese to carry on their dietary tradition of "no such thing as too much refinement in food."

Food and Health

Imbedded into the Chinese's food culture is an extremely important backbone—therapy by diet. The tradition of the cross-use of foods and herbal medicine has existed since archaic times. The god of agriculture in Chinese legends, Shennong, not only taught people to grow crops, was also the master of medicine having "tasted hundreds of kinds of herbs." Although it is a mere legend, it reflects an important idea in traditional Chinese medicine—"food and medicine share the same roots." So people's dietary intake and disease prevention and treatment have quite a tight connection.

The Chinese since days of old have paid close attention to the preservation of health and prolonging of life. The book of *Huangdi Neijing*, first introduced a dialectic and comprehensive view on diet. Only when the diet is comprehensive can nutrition be complete and balanced. The "five tastes" should also be all included in the diet so that no particular taste would be excessive enough to hurt the internal organs. Relying on daily diet to improve physique and health and fighting sickness is the important essence of Chinese food culture. Compared with medicine, food is more gentle to the body. Every type of food contains certain "fine extract" that can exert different effects on the body. For the purpose of relieving excess heat in the body, Chinese doctors believe pears work for the lungs, bananas help the rectum, while kiwis do the same but for the bladder. Different tastes also have different influence on the body. Usually, sourness goes to the liver, pungency goes to the lungs, bitterness into the heart, saltiness into the kidneys and sweetness goes to the spleen. Different elements are absorbed by different internal organs, and have different effects on the body. Relying on the different nature and nutritional contents of foods to influence the physical body is a unique feature of Chinese food culture.

Hot and spicy foods can facilitate the flowing of *qi* (chi) in the five organs. In summer, with high humidity and heat, beverages such as mung bean soup, sweet-sour plum juice, lily bulb soup, chilled tea and so on, are great at protecting people from heat fever. In the dry autumn air, it would be beneficial to consume foods that moisten the lungs, such as pear, persimmon, olive, turnip, and tremella. Common Chinese folks prefer turnip since it is inexpensive and has visible health-improving effects. Turnip with braised ribs or with simmered lamb all have tonic effects. Chinese chestnut, Chinese yam and river snails are also great health foods for the fall. Winter is the best time for "tonic intake." The Chinese

A shop selling shark fins in Guangzhou. (Photo by Jie Zhu, provided by image library of Hong Kong *China Tourism*)

Ginseng is an excellent tonic. (Photo by Gang Feng)

upon entering the winter season, like to have chicken, pig feet, beef, lamb, longan, walnuts, sesame and other high fat and high calorie foods. Especially for those who have frail bodies and are afraid of the cold, a little dog meat can help replenish heat in the body.

People of different ages have different ways of therapy with food. Middle-aged people experience a time when the body slowly turns from vigorous to weak. They need high-energy foods with protective properties as well as age-defying foods to slow down the aging process. With their slow metabolism, middle-aged people should eat four-legged animals less, such as beef and pork. Instead, they should have more "two-legged" animals, referring to poultry, or the "one-legged" fungi and the fish, with "no leg."

Food therapy among common people in China has long been popular. There is a saying that "medicine falls short when compared to food in providing supplements to the body and food works just as well as medicine in treating illnesses." The fact that common fruits and vegetables can prevent and cure sickness is

Depends on the season, great differences exists when taking tonic foods. (Photo by Gang Feng)

known by every family. When a family member comes down with a cold, cut a few ginger slices and add a few pieces of scallion. Then add brown sugar to boil in water and drink while hot. Finally, sleep under some thick blankets to induce loads of sweat and the cold is gone. Hen stewed in clear soup, millet with brown sugar and sesame seeds that have been stir-fried are the first choice for women after labor. It helps them quickly restore bodily energy and functions, relieve excessive heat and revitalizes.

Medicinal diets are made by mixing traditional Chinese medicine with conventional food and cooking them together. The variety and dosage are strictly controlled. It is different from using food in place of medicine, since medicinal diets are mainly to make unpleasant-tasting medicinal herbs taste like delicious foods for easier taking. Taking medicine can then be as easy as having a meal. Such combination of medicine and meal forms a new breed of food by taking properties of medicine and the taste of great food. This also pushed Chinese therapy by food to a new level. Popular medicinal diets include porridge, pastry, stews and dishes such as

Duck with Chinese caterpillar Fungus, Whole Chicken Stewed with Ginkgo Nut, Stir-fried River Snails with Rice Wine, Pig Stomach with Lotus Seeds, lily bulb porridge, *fuling* (an edible fungus) cake, Chinese yam and more. Today, China's cities both large and small have specialty medicinal diet restaurants, and business is rather prosperous.

Having pig feet regularly can strengthen the body and is helpful to slow down aging, keeping a smooth skin. (Photo by Gang Feng)

Chinese medicinal diets not only shone brightly domestically, but are also being introduced overseas. It is gaining wide acceptance and imitated by foreigners, and is becoming part of their local food culture. Chrysanthemum wine, wine made from bark of slender acanthopanax, ginseng wine, oolong tea, ginger juice candy, sour plum and other traditional Chinese tonic drinks and foods all have a large market in foreign countries. Gin, the popular Western drink has its main ingredient being a Chinese medicinal herb—the seed of Oriental arborvitae. It has the ability to calm and relax people.

Casserole Black Chicken stewed with Chinese herbal medicine. (Photo by Jie Zhu, provided by image library of Hong Kong China Tourism)

Chinese food therapy and medicinal diet practice now gains more and more westerner adopters. What this represents is the unified desire for health and long age. Though Western medicine can cure many pains and illnesses, but due to its properties of chemical compounds, it causes rather strong side effects. Needless to say that it has no nutritional value. Chinese medicinal diets, however, are fully rooted in natural plants and herbs; long-term use at the right dosage is much safer. Even more important is that it can nourish the body and preserve health by strengthening the body's immune system, achieving the goal of slow-aging and prolonging life.

When speaking of taste, it usually represents the unique characteristic of various regional cuisines. But

People believe that having fish head regularly can help slow down the aging process. (Photo by Gang Feng)

from a health preservation standpoint, diets heavy on saltiness, sweetness, sourness or pungency are all unfavorable to the body. Consuming too much salt will damage the heart, spleen and kidneys. Too much sour and pungent taste will cause various forms of ulcers. Therefore, the way to long-lasting health is through harmonious balance of the "five tastes" and going lightly on the seasoning.

Before the present day, the Chinese ate much less meat. Not only because economic conditions of the past did not allow having much meat, the Chinese also saw meats as supplements to crops and vegetables, usually having meats and non-meats in combination; whereas Westerners place meats and the like as the main component of their dietary structure. Meats are not the principal part in the dietary make-up of the Chinese for health and nutritional reasons with sound scientific basis.

The development of vegetarian dishes and the spread of Buddhism are intimately related. When Buddhism was first introduced to China, there were no strict abstentions on food.

Later, in the Southern Dynasties (420—589 AD), the devout Buddhist, Emperor Liang Wudi (ruled from 502 to 549 AD) believed that eating meat equaled the act of killing, so was against the Buddha's teachings. As a result, Buddhist temples begin to forbid the taking of wine and meats. The monks ate vegetarian food all year round, even as to influence the lay Buddhists, those who stay at home, working to achieve enlightenment. The increase in number of vegetarians hastened the development of vegetarian cuisine. All the way until the Song Dynasty, literati and men of great accomplishment promoted vegetarian dishes, thus it was able to show the world all its glory. Bean curd, gluten, and vegetables were the main ingredients to vegetarian food, and they gradually became real delicacies in people's eyes. The food industries of the common people also begin to develop and market vegetarian foods to satisfy the needs of the Buddhists, and also influenced the vegetarian food varieties of the monasteries. Because vegetarian dishes usually have little taste, so it must be skillfully cooked for the general public to accept. Only then can it be on par with tantalizing traditional delicacies.

The Chinese since ancient times have had the practice of eating porridge, or congee, to prolong life. The usual way is to have one bowl of thin porridge on an empty stomach every morning. Porridge prevents illnesses and preserves health. Chinese people have long since proven through practice that carrot porridge can prevent high-blood pressure. Those who are accustomed to excessive meats and seafood can have some vegetable or wild herb porridge to increase essential vitamins, and benefit the kidneys, which are organs with the nature of Yin. Having less meat and seafood, but more vegetarian dishes and porridge have always been the second-to-none choice for people looking for health. Vegetarian dishes with vegetables, fungi, and bean-based ingredients, are easy on digestion, and are nutrient-rich. Proven by modern medical science, vegetarian food is the healthy way to eat and deserves to be widely practiced. But eating only non-meats is also not appropriate, as it is not balanced and well rounded enough in nutrition. Such as the essential element of the body—Calcium, which is rarely found in vegetarian food. Therefore, back to the philosophy of "five tastes in harmony" and balanced diet, we find that a reasonable dietary structure becomes evermore dependent on dietetics and health studies, making people more aware of nutritional health.

Food Taboos

Chinese philosophy emphasizes "the nature and the people as one." This kind of cultural mentality is reflected in the diet, by way of the food and the eater's harmonious co-existence and growth. Therefore, there came to be many taboos in the daily Chinese diet. These include the reasonable combination of food, seasonal or daily taboos, as well as "catalysts" and foods not appropriate during times of illness. Some of these prohibitions were passed down the

Foods made from sticky-rice have a great texture, but are not easy on digestion. People really love it but dare not have too much. (Photo provided by Xinhua News Agency photo department)

generations and discovered from experience, others were learned from scientific findings of modern society. In short, the problem with eating is not that simple.

The Chinese focus on the mixing and combination of food. Dumplings go with vinegar; scallions wrapped with pancake use dips; when having deep-fried twisted dough sticks, one must also have soy mean milk; and noodles cannot be enjoyed without toppings and dressings. A table full of home-cooked meal must have both meats and non-meat dishes, balancing the yin and the yang. Most importantly is the combination of principal foods and supplementary foods, such as rice and beef, where beef tastes sweet and cool, and rice is slightly bitter and warm. The two items are offset when sweetness is couple with bitterness, a truly excellent combination of principal and supplement. In addition, lamb and glutinous millet; dog meat and sorghum; pork and millet; and poultry (birds) and dough are all great combinations. In contrary, some foods do not add up well. If these are mixed, it will spell trouble and harm one's health. Some examples include plums and white honey, as it will attack the internal organs; soybean and pork cannot be had together; neither can mustard and rabbit meat and so on. These are all taboos in mixing foods. Many people have had this kind of experience, when after a hearty banquet, they do not feel any satisfaction, instead, feel an unexplainable discomfort, or even falling victim to sickness. This is because they have had too much and too many types of food all at once, including foods that are opposing in properties.

The food taboos among the Chinese general public have much to do with seasonal characteristics. Meaning that diet must change as the seasons change. The same kind of food is fitting for a certain time but not for the rest. This is "seasonal taboos in food." The general public believes that having leek in winter and spring can "warm the back and the knee." However, in summer, leek makes people "dizzy with poor vision." Dog meat is a real tasty treat during wintertime, but should be avoided in other seasons. Fresh

hot peppers are loved by the people in Jiangxi Province in the summer. For the winter, dried hot peppers can still be had. But for autumn, basically no hot peppers are in the daily diet.

From daily diets, the Chinese learned about and discovered many "no-no's." Such as breakfast cannot be dry foods or eggs only; no smoking during meals; no watching TV when eating; and never get angry during a meal. Before and after meals, drinking too much water or having cold drinks are not recommended; strong tea and fruits are also advised against. After laboring one's vocals, cold drinks are damaging to the throat. Before going on trips and riding on motor vehicles, avoid having a full stomach. After physical exercise, too much sugar would also be inappropriate. The list goes on and on.

For some people, these taboos to daily food intake are not worthy of concern. However, as soon as one feels discomfort or it is during special times such as pregnancy, these taboos must not be taken lightly. As the common saying goes "thirty percent treatment, seventy percent prevention." If one does not understand the

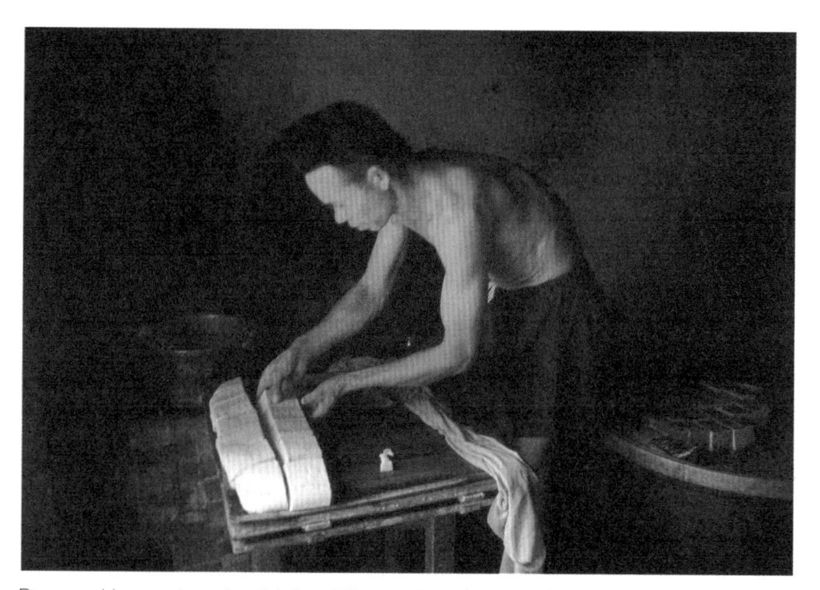

Bean curd is easy to make, rich in nutrition, and can be made into many varieties of dishes. (Photo by Zhenge Peng, provided by image library of Hong Kong *China Tourism*)

< Muslims of the Uyghur nationality enjoying Hand-Served Lamb at the wedding banquet of friends and family. (Photo by Miao Wang, provided by image library of Hong Kong *China Tourism*)

"catalysts" or do not avoid the "tabooed foods," it is very possible to experience negative reactions in the body, or even serious illness.

The so-called "catalysts" are foods that could spur sickness. The range is broad including heads of chickens, heads of pigs, seafood, fish, beef and lamb, as well as various types of spices and seasoning. Depending on the condition of the body and the type of illness suffered from, the types of "tabooed foods" change accordingly. If one is weak and feeling cold in all four limbs, one shall not have watermelon, banana and pear, which have a chilling effect. If one suffers from excessive body heat and thirst, amnesia or anxiety, it is best not to eat ginger, black pepper, rice wine and so on. When asthma attacks, eggs, milk, fish and seafood with high protein become "tabooed foods." After suffering a cold, one should keep away from chilled drinks and cold foods, greasy and thick-tasting foods, or highly pungent foods. When taking tonic, tea and turnip should be avoided, or it will comprise the effect of the tonic.

For women who are expecting a child, the emphasis in their diet is on all-round nutrition, with no excess of anything. They should eat little or none of foods that are pungent, hot, warm and dry, or oily and indigestible. In the three or four days after giving birth, new mothers should stick to a vegetable diet, since eating meat and fish, especially 'enhancing' foods like carp, is very unhelpful to the healing of post partum wounds, whereas warming tonic foods like mullet are very good for that healing process. Later, crucian carp, pig's trotters, eggs, and other 'enhancing' foods are often taken to promote lactation.

As for women during pregnancy, the main principle to follow is to have all-around nutrition with no particulars to certain types of food. But due to required pre-natal environment for the baby's growth, a pregnant woman's body is usually strong in the qi (chi) of Yang. So hot and pungent, greasy, and hard-to-digest foods should be restrained from. After labor, food such as crucian carp, pig's feet and eggs are often incorporated into a woman's diet to induce milk secretion. But in the few days immediately after giving birth, a vegetarian diet is recommended; meats and especially carp and crucian carp are not favorable to the healing of cuts. However, the tortoise which can warm the body and replenish qi (chi) is great for healing of cuts.

Vegetable shelves in a supermarket. (Photo by Ran Jing, provided by Imagechina)

Of course, some of the abstentions in foods for the Chinese is only a custom among the common folks, and have no real scientific base. For example, certain places in China believe that a pregnant woman cannot eat rabbit meat or the child will be born with three pieces of lips like the bunny. She also cannot eat donkey meat or the child's face will be as long as the donkey. Tortoise meat, eel and loach are also believed to give the baby small head, face and eyes respectively. Looking at these beliefs with scientific knowledge, it is obvious that these taboos in foods are simply nonsense. However, it holds the people's wonderful hopes for the next generation, and can also be considered a reflection of Chinese folk culture.

So it seems that the study of diet is an extensive area of knowledge. Diet and culture have a definite link. The favored food of a particular culture may be frowned upon in another culture. The Indians worship the cow as a sacred animal and laws ban the killing of cows; Jewish people detest and reject pork; but the Westerners have no objections to neither of these two kinds of meat whatsoever. But when Westerners see some nationalities feasting on insects or dog meat, disgust rushes forth. Therefore, taboos in diet can also be seen as abstention culture in food culture. But abstention culture, aside from reflections in daily life, is also tightly intertwined with religious faiths, nationalities or customs and traditions of a particular industry in a particular country or area.

Sweet potatoes are traditional agricultural crops planted widely in China. The total yield in China amounts to 80% of worldwide yields, but eating sweet potatoes directly only occupies a small portion of modern Chinese daily diet. The picture shows farmers harvesting sweet potatoes near the Yellow River in Shaanxi Province. (Photo by Wugong Hu, provided by image library of Hong Kong *China Tourism*)

Chinese food taboos in religion are usually reflected in Buddhism, Daoism and Islam. For example, Chinese Buddhists are forbidden to eat meat, as eating meat is considered one and the same as killing of life, a defile of religious discipline. However, in India, Sri Lanka and other countries, as well as the Mongolian, Tibetan and the Dai minority nationalities of China, the monks do not follow such a religious commandment. Actually, it is more in line with ancient Chinese courtesies during sacrificial rituals, when people bath and change clothes, but do not eat meat or drink wine in order to show devout faith. So apparently, what forbids Buddhist monks from eating meat or drinking alcohol is more due to traditions and customs of the Han Chinese culture. The Daoist faith, which is indigenous to China, also refrains from eating meat and drinking alcohol. However, different from the Buddhist approach, Daoists merely focus on preserving the inner organs and nourishing the spirit. They do not advocate complicated taste in food, since immortals are supposed to have become who they are because they do not eat earthly foods. Islam forbids its

An age-old traditional Chinese pharmacy. (Photo by Guanghui Xie, provided by image library of Hong Kong *China Tourism*)

believers from eating animals that died naturally, or having blood and pork, as well as animals that died by strangulation, falling, or the leftovers of other carnivorous animals. Muslims consider the aforementioned categories of animals "unclean." However, if the animal, aside from pork, were butchered, it should be fine for consumption. Aside from these animals, domesticated animals such as horses, mules and donkeys, plus "scale-less fish" such as crabs and eels are all on the tabooed list of foods.

Moreover, folk traditions and local customs as well as certain professions and industries have also contributed to abstentions in diet.

Certain southern regions of China happily enjoy wild animal meats and seafood. However, eating snakes, a common practice in these areas is considered an act of sacrilege in other places, where the snake is believed to be the guardian angel of humans. It should instead be loved and cared for so that man and snake can live in peace and harmony.

Fishermen living along the coast have also many taboos in food due to their special professions. Their main dish is obviously fish. At the first meal of fish during a new year, the raw fish must be taken to the bow of the boat, to be offered as sacrifice to the dragon king and sea god. When eating the fish, after finishing the meat on the side of the fish that is facing up, the complete carcass must be taken out before eating the underside; the fish must not be turned over. For each meal, there must be leftovers of the fish, with a bowl of fish soup, to be poured into the pot for cooking of the next meal. The significance of these customs is to symbolize the constant supply of fish. The leftovers of meals, including the carcass of the fish and the muddy waters after washing the dishes, shall not be dumped into the sea.

The Tibetans at China's western frontier have many taboos in meats. Snakes and aquatic animals are rarely eaten. Some people do not even eat eggs, never mind birds. The Tibetans never hunt for food, especially the snow pigeon, which is worshipped as a sacred being. Even for beef or lamb, they never eat fresh meat from livestock killed that same day. It is believed that although the animals have been slaughtered, their souls still remain. So people must wait until the second day before consuming the meat. Garlic is also a taboo for the Tibetans. When worshipping on sacred grounds, garlic can taint the holy place.

Comparing the kinds of taboos to food and diet between different nationalities and areas can be a fun thing to do. Different nationalities can have totally opposite abstentions in food. The Miao people forbid the killing of dogs and eating its meat, but dog meat is the favorite food of the Koreans and is often used to treat guests. At banquets of the Miaos, chicken and ducks are often used to treat guests; chicken hearts and livers being the most precious parts and are offered to the elders or the guests first. However, the Nu nationality abstains from killing chickens to treat guests, so never mention the wish to eat chicken when visiting a Nu family.

Some nationalities have abstentions of diet that are simply absurd to other people. For instance, the Yi nationality of Yunnan Province, have not just numerous, but strange taboos. If the chopsticks break while one stirs the food with it, then the food is no longer edible. One also cannot make food from the flour if the mill spindle breaks during the milling process. If a sheep suddenly cries out right before it is to be butchered, then its life will be spared. When a table has just been set with full of food, if a chicken accidentally hops over the dishes, then the meal must be remade. Children cannot eat chicken stomach and tail, pork ears, sheep ears and so on.

When looking from the viewpoint of folk or regional customs, there is no right or wrong, better or worse with abstentions in diets. The interesting thing is, abstentions in diet inadvertently created a line-up of quite zesty and unique foods and drinks. Chinese specialty vegetarian dishes are the most typical of this group. Many Buddhist temples and Daoist monasteries have their fine specialty vegetarian fares, with fresh taste, elegant appearance, great variety, colors and shapes that all rival meat dishes. Fa Yuan Temple of

The Chinese who believe that "food and medicine share the same roots" prefer tonic foods. (Photo by Gang Feng)

Beijing has the Koumo (a kind of dried mushroom) pot stickers; Nanjing's Bao En Temple has Soft Fragrant Cakes; Also in Nanjing, "Cow Head" Tofu by the Monk Xiao Tang is quite popular; and in the city of Xiamen, there is the Soup Vegetables by the South Pu Tuo temple. All of the aforementioned dishes are the special pride of the various holy places.

A Gastronomic Tour of China

All About the Names of Dishes
On a Chinese dining table, there is no dish without a name. A great number of these names come from some aspect of the food or its preparation, for example 'Mixed Vegetables' or 'Chub Beancurd' from the ingredients, 'Fish-fragrant Slivers of Pork' or 'Hot and Numbing Fragrant Pot' from the flavour, 'Taichi Beancurd' and 'Squirrel Mandarin Fish' from the shapes, 'Fragrant and Crisp Duck' and 'Soft-fried Tenderloin' from the texture, 'Pearl and Jade Soup' 'Gold and Jade Chowder' from the appearance, 'Plain Scalded Green Vegetable', 'Meat Steamed in Flour', 'Dry Fried Eel' and 'Salt Steamed Chicken' from the cooking method, 'Deep Fried Autumn Pancake' from the season, 'Four Delight Meatballs', 'Eight Treasure Beancurd Pot', 'Nine Turn Intestines' and 'Mille-feuille Pancake' from some number association, and 'Mapo Beancurd' and 'Dongpo Pork Legs' from a person's name.

Particularly fine foods are called by the Chinese 'treasures of the mountains and flavours of the deep'. Historical records show that among the rare and precious delicacies once selected for the menus of the great and powerful were such things as bears' paws, swallows' nests, sharks' fins, sea cucumber, elephant trunks, camel humps, deer tails, and monkey brains. But foods like this have become a complete rarity on the banquet menus of modern China, and furthermore, with the increasing awareness of the need to cherish and protect the animal world, many people have chosen to renounce foods of this kind. If you want a genuine example of how eating trends in China change with the times, you should look to the distinctive and influential dishes of the regional cuisines.

The Chinese are accustomed to calling all kinds of delicious fare "rare tastes from the mountains and seas." These delicacies include bear paws, bird's nest, Fish Wing, sea cucumbers, elephant trunk, camel's hump, deer tail, monkey brain and the like. These high-class foods are now very rare in Chinese diet, even in high-class restaurants. With increased awareness for wild animal preservation in recent years, guiding the trend of food culture, instead, are some very unique regional specialties.

The duck, in traditional Chinese medicine, is a premium tonic food. The Chinese believe eating ducks regularly is beneficial to health. (Photo by Gang Feng)

China's distinguished geological and climatic conditions and resources in its many regions, together with the unique dietary habits of the local people, have forged the many different styles of Chinese cuisines. Such as Lu, Yue, Su, Jing, Min, Zhe, Xiang, Hui and more. The common people of China generalizes the different regional tastes as "South is sweet, north is salty; east is Spicy and west is sour."

As a global metropolitan, the historical capital city

of Beijing has thousands of restaurants, with at least one hundred famous ones. These restaurants not only incorporate various Chinese regional food styles, but also Western tastes such as authentic French, Italian, Russian, Spanish and American foods; and other Asian cuisines including Japanese, Korean, Indian, Vietnamese, Indonesian, Thai and more. In recent years, with consumer spending power on the rise, Beijing is already marketing several distinguished food streets. And more and more 24-hour service restaurants are appearing. When it comes to timeless favorites of Beijing, one cannot miss the Beijing (Peking) Roast Duck and Lamb Hotpot (fondue). A famous food house, Quanjude, roasts their ducks with open fire, while another well-known restaurant, Bianyifang, uses enclosed fires. Each of the two styles has its unique strong points. A glistening brown roast duck, its meat sliced, dipped into sweet noodle sauce, with the choice of some scallion threads, rolled into one soft, thin dough wrap. The taste and feel of the duck wrap inside your mouth is the most satisfying. The Lamb Hotpot used to be a winter dish, but is now available anytime as most restaurants have air-conditioning equipped. Even in the seething summer heat, many people prefer it. Close friends and family sit around a round table having hotpot, and order a few plates of raw fresh lamb and veal slices. On top of that, add some seasonal greens and boil it all in a bubbling hotpot. When the meat is cooked, bring it out and enjoy with a variety of sauces including sesame oil dip, tofu with paste, flowers of Chinese chive, or hot pepper oil, and sprinkle with some minced scallion and Chinese parsley powder. When you have had enough meats, cook some super thin

Several years ago, the Lamb Hotpot was still a premium treat for guests. The host and guests sit around the hotpot, conversing freely while savoring the delicious food. Aroma of the meat is most satisfying and people feel a strong sense of affection. (Photo by Roy Dang, provided by Imagechina)

The Beijing Tan Family Dishes that attach great importance to color, smell, taste, shape and supplementary ingredients. (Photo by Jianhui Zhu, provided by image library of Hong Kong *China Tourism*)

Lotus roots are a specialty food item of China. (Photo by Xiaofei Wang, provided by image library of Hong Kong *China Tourism*)

rice noodles. The rice noodles, after absorbing the broth from the pot, have excellent taste and texture. For some really unforgettable experience, have one or two small sesame seed cakes afterwards.

Beijing's neighbor, Tianjin, is a famous port city. Its cuisines are also typical of the northern style. Its most famed food, the Dog-Won't-Eat Stuffed Buns, has the characteristics of thick-tasting brine (extract) and symmetrical wrinkles on the bun as a result of kneading the dough, sealing in the stuffing. It is said that each bun cannot have less than 15 wrinkles. The Tianjin Sesame Twist is another famed local specialty. It is crunchy and with wonderfully appetizing aroma. Especially the 18-Street Grand Twists made by historic franchise food house—Guifaxiang, are the most renowned. Tianjin, as a port city, has plenty of fish, shrimps and crabs. Usually these freshwater and saltwater resources are the principal ingredients in Tianjin-style cuisines. Its cooking techniques mainly involve boiling and simmering. Most people know of Tianjin food as inexpensive and having good value for large servings, attracting visitors from all around China and the world. In addition, in 1860, the city of Tianjin was selected by the Qing (Manchu) government to become an open trading port; thus Western food has taken root here. A German chef established here, Qishilin, the famous food house now with over 100 years of history. It is famous for selling authentic German, French cuisines and Western pastry and cakes.

Going a little further to the southeast, one would arrive in Shandong province. Due to the early development of its regional cuisine, the Lu style is one of the most influential and popular regional food styles in China. Shandong is home to the great philosopher Confucius, therefore, its food style is a true embodiment of Confucian teachings—"No such thing as too much refinement." It emphasizes purity of seasoning, and is salty and delicious; with main characteristics of natural flavoring, tender, fragrant and crispy. Commonly used Lu techniques exceed 30 in number; especially excellent are its quick-fried, stir-fried, roasted, and braised dishes. In the Ming and Qing dynasties, Lu cuisine is already the main component of imperial diet. The state banquet of the Qing Dynasty, Manchu-Han Full Banquet, uses all real silverware, and has a total of 196 courses that are all real delicacies; extravagant to the extreme. As the top cooking style of northern China, Lu style cuisine is the blue print from which the basic dishes of high-class and festive banquets, and home style cooking are developed. Not only this, the Lu style also heavily influenced the regional foods of Beijing, Tianjin and northeastern China. Worthy of mentioning is the Fushan area of the Jiaodong region, it is known throughout the world for its high development of culinary arts. There, famous chefs of all ages abound. Not only are its chefs armed with the most excellent cooking skills, even the "head chefs" in each household can make excellent fare. Overseas Chinese originally from Fushan have spread the Lu style to other places throughout the world. Shandong people are known for their warm hospitality. They are afraid to see honorable quests not being fully indulged with excellent taste, and with stomachs half-empty when meals are finished; so Shandong dishes come in very large servings. Those who visit Shandong must be ready to eat like they have never eaten before.

A close neighbor of Shandong is Shanxi Province. Though the food in this province was never considered one of China's several main styles, it does not prevent the people here from enjoying exquisite tastes. The Shanxi people have a long tradition of being merchants, especially during the Ming and Qing dynasties when trading flourished. There emerged many wealthy merchants and made Shanxi the "richest place in the country." Therefore, if one just pays a bit more attention to Shanxi's menu, it is not difficult to find evidence that ultra-wealthy people used to live here. Just for noodles, there would be countless unique ways of preparation, each one being different from the other, and even as to have a grand feast with just noodles.

Just as the people of Shanxi like to eat flour-based food, the people of Henan also give flour the leading role. The most strikingly original feature of Henan cuisine is the importance of soup. The famous 'Liquid Banquet of Luoyang' consists of 24 dishes meticulously prepared, 8 cold and 16 hot, some of them vegetables and some meat and fish, and there is soup in all of them.

West of Shanxi lies the province of Shaanxi. The provincial capital is the ancient city of Xi'an. Aside from housing the timeless terracotta warriors, the Great Wild Goose Pagoda and other historical sites, two things that can attract outsiders to Xi'an are none other than Soaked Buns in Mutton Soup and dumpling feasts. From large avenues to small streets, countless restaurants specialize in Soaked Buns in Mutton Soup. The customers would, with their own hands, tear the steamed buns into pieces, the smaller the pieces the better. Then hand the bowl of buns to the head chef inside the kitchen where he would drench the buns with

The snack street of Wufujing in Beijing. (Photo by Wei Deng, provided by image library of Hong Kong *China Tourism*)

delicious mutton soup before serving. Dumplings are the northern Chinese's traditional food. The dumpling feast of Xi'an consists of steamed, boiled, deep-fried and fried dumplings, with as many as 108 different varieties. The dumplings have fine supplementary ingredients and unique shapes. Some look like butterflies or bird's nests, others resemble seashells or even clouds; all look and taste different. When having dumplings, one can listen to many folk legends or historical facts related to dumplings; it is an experience that is one of a kind.

Beautifully shaped soup-filled buns. (Photo by Qitao Yang, provided by Imagechina)

Going even further west from Xi'an, at Yinchuan, people can enjoy authentic roast lamb's head; at Lanzhou, there's authentic Beef and Stretched Noodles; Xining has Lamb Innards Soup that should not be missed; and in Ürümuqi there are sticks after sticks of lamb kabobs waiting. From Xi'an to the north, deep into the grasslands of Inner Mongolia, one must have a taste of the roast whole lamb.

From Xi'an to the south, one would arrive in "the Heavenly State"—Sichuan. Sichuan food is also an alternative regional style that has matured long ago. It poses great influence over all regions of China. When people think of Sichuan food, almost the only thing that comes to mind is the pungent and hot flavor. In actuality, Sichuan food pays great attention to flavor and seasoning, with tastes that one can see. From the seasoning, we see scallion, ginger, garlic, hot pepper, black pepper, star aniseed, vinegar, thick broad-bean sauce, fermented rice soup, sugar, salt and many more. As long as attentive efforts are made, Sichuan food can have as many as seven flavors, which are sour, sweet, bitter, pungent, mouth-numbing, fragrant and salty. Most Sichuan dishes are popular home-style cooking,

Fragrant Skewers in Chengdu. (Photo by Hongjiang Zhang)

with characteristics of being simple and refreshing. Many people who have been to Sichuan say the great foods of Sichuan are countless, including home-style Yuxiang Pork (pork in fragrant spicy sauce), Twice-Cooked Pork, Bean Jelly in Chili Oil, Mapo *Tofu* (hot and spicy bean curd), *Fu Qi Feipian* (literally "married couple's slices of lung," made with spicy beef and beef lung/stomach/tongue slices) to street-side snacks including Fragrant Skewers, Numb-hot (*mala*) Rabbit Head and Noodles in Chili Oil (Dan Dan Noodle); plus the all-famous Chinese dish of Numb-hot (*mala*) Hotpot and Boiled Fish in Xhili Oil, all are foods that one can never get enough of.

When it comes to hot flavors, all provinces and regions in western China have the tradition of eating hot foods. It is commonly perceived that hot foods have the effect of waning off cold and wetness. Hot peppers were introduced to China from the Americas at the end of the Ming Dynasty. In the beginning, it was only used as a decorative or medicinal crop. The first regions to eat hot peppers as food are Guizhou Province and its neighboring areas. At one time, hot peppers were used in place of salt as a seasoning. In the present day, not only Sichuan food is famed throughout the country for its hot and pungent flavor, those neighboring provinces including Shaanxi, Guizhou, Yunnan, Hubei; and south central provinces of Hunan, Jiangxi and Guangxi all have different styles of hot and spicy food; as each area has one other main feature together with pungency. Sichuan focuses on *ma*, or mouth-numbing taste; Guizhou focuses on fragrance; Yunnan focuses on taste of natural essence; Shaanxi focuses on saltiness; and Hunan focuses on sourness. In recent years, as regional foods make their way into large metropolitan cities such as Beijing,

Chongqing *Mala* Hotpot (Photo by Hongjiang Zhang)

Shanghai, Shenzhen, and Guangzhou, regional styles including Hunan, Hubei, Guizhou, and Yunnan have met with acclaim from more and more eaters.

Xiang style, short for Hunan Food, is one of the eight main food styles of China, and has certain popularity on the international scene. Xiang food is known for having fine knife work, being heavy in oil and strong flavored. It mostly uses techniques such as boil, roast, and steam to make hot and sour; charred and mouth-numbing; fresh and fragrant; crispy and tender; smoked and cured; as well as dishes in many other flavors. Hubei food is famous for its refined and meticulous process. A single dish often has to go through more than a dozen steps in the making. Hubei food uses mainly aquatic products as its ingredient. Steamed food is Hubei's strong point. It has the characteristics of thick broth, full and pure-flavored. Guizhou food is known for its many rare delicacies made with wild land animals as well as chicken, duck, pork, beef, vegetables and bean curd, with principal flavors of saltiness, pungency and fragrance. Integrating culinary techniques of local minority nationalities, Guizhou food has a very rustic twist to it. Famous dishes include Chicken in Dry Hotpot, Fish in Sour Soup, Huajiang Dog Meat and more. Yunnan is a province with high concentration of minority nationalities. Its food style is distinctively local. Different types of fungi are the regional specialties as the area is full of all types of edible wild fungi. Guangxi food excels in making wild animal and plant dishes, and focuses on natural and fresh flavors. It is influenced by Cantonese cuisines but is also into hot foods. There exist local minority nationality characteristics in Guangxi food. Being an area rich in the yield of precious Chinese medicine ingredients, Guangxi people artfully mix food and herbs together to make tonic meals as one of the specialties of the area.

When the subject of Cantonese food comes up, people always want to know why Guangdong Province is more particular about food than any other province in China, and also why the cuisine of the province, with Cantonese cuisine as the leading example, has

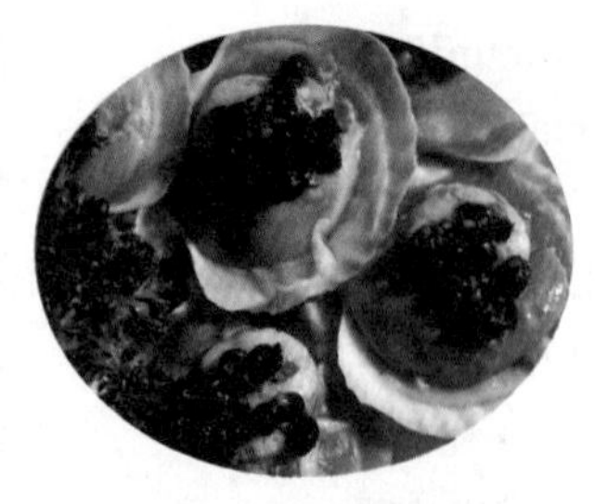

Yue cuisine from Guangdong is known for a whole array of ingredients and the creativity and uniqueness of its dishes. It is also deeply concerned with nutrition. (Photo by Gang Feng)

uniquely combined and blended the food traditions of so many other places. Guangzhou is located at the delta of the Zhujiang River. It has convenient water routes in all directions thus has been the southern commercial center of China. In addition, Guangzhou is China's earliest port city to be opened to foreign trade. All kinds of traveling merchants from all parts of China brought with them their regional restaurants, with a multitude of tastes and styles. Aided further by rich local natural resources, fresh seafood and rare land animals are all possible dishes on the dining table. In cooking, Guangzhou food takes the strong points of Western cuisines; making it the most extraordinary among Chinese cuisines, with characteristics of an extensive list of ingredients, fancy presentation and being nutrition-conscious. The Cantonese love to eat, and are health-conscious, famous for making seasonal soups and porridge. Fujian, which borders Guangdong, has a food style very different from its neighbor. With Fuzhou food as representative of the Fujian style, the mildly seasoned dishes use fresh ingredients, bordering on sweet and sour in taste, and come in mostly soup dishes. The Fujian people are skilled in making seafood with a twist, and the use of distiller's grain and wine in foods. Its most famous dishes include "Buddha Jump Over the Wall (simmered mix of seafood and land delicacies in soup)," Fish Balls in Clear Soup, Sea Clams Quick-boiled in Chicken Soup, Chicken Dice in Distiller's Grains and much more.

A well-known dish in China, the *Dazha* Crabs from Yangcheng Lake. (Photo by Guanghui Xie, provided by image library of Hong Kong *China Tourism*)

The provinces of Jiangsu, Zhejiang, and the city of Shanghai, located on China's east coast, have always had close geographical relationships. Thus their food cultures pose reciprocal influence on one another.

Due to the relatively short history of the city of Shanghai, its foods all originated from Ningbo, Yangzhou, Suzhou or Wuxi, even with influences from Sichuan. The real historical local specialties are the Yangzhou, Suzhou and Wuxi styles in the province of Jiangsu, as well as Ningbo and Hangzhou styles from the province of Zhejiang. The one characteristic of Yangzhou cuisine is that it strives to lock in the natural flavor and juices of the ingredient. So different dishes come in different flavors. Furthermore, Yangzhou's snack foods are famous near and far for its great variety. Suzhou, on the other hand, is a historical town deeply rooted in humanistic culture, being once the home to many of the famed literati and great talents. Suzhou cuisine's relentless pursuit for refinement is also well-known. It emphasizes skilled application of cutting, boiling, supplementary ingredients and seasonings, and most importantly the duration and degree of heating. Even a round of home meal would get the utmost attention to refinement, focusing on quality, not quantity, and tastes light and refreshing. In recent years, the Dazha Crab from Yangcheng Lake, which started a dining craze all over the country, is also a native animal to Suzhou. Wuxi food

A Yellow River Carp banquet from Henan Province. (Photo by Zhixiong Li, provided by image library of Hong Kong *China Tourism*)

has two distinct characters, "sweetness" and "stench." Almost all homes use crystal sugar powder, thus the sweet taste. For the "stench," Stinky Tofu is used. But the more odorous the bean curd, the better the flavor and aftertaste. As experienced eaters believe, Wuxi food is among the most excellent in both knife work and degree of heating. Ningbo food relies on the high production of local seafood as main ingredient, usually being a bit on the salty side. Hanzhou is another city with over one thousand years of history. Not only are its sceneries most enchanting, its fine foods are well worth mentioning. The people of Hangzhou are proud of the rate at which they overthrow the old-style cuisines and creating new tastes, which is record-breaking in China. Hangzhou food is also mild and refreshing, with next-to-nothing usage of pungent spices and thick sauces and oils. The only exceptions being the Dong Po Pork Leg and West Lake Sweet and Sour Fish, which are hailed as classic fare, with an lasting fragrance long after finished eating.

People cannot help but be reminded of Anhui food, which took the country by storm one hundred years ago. It is said that the Anhui-style restaurants at the time were all very large in scale, with full redwood interior showing an air of wealth and loftiness. But in the modern race of restaurants in the food service industry, Anhui food faded away from the spotlight. If it were not for tours to the Huangshan Mountains, outsiders can hardly ever get a taste of authentic Anhui food.

All around the country, from snacks to full feasts, China has an uncountable list of great cuisines in each and every region. Exotic fare in all forms and flavors project the splendor of a long tradition of food culture, with the vivid local culture of the regions for the respective cuisine styles. To taste all of China's foods is not only a long and luxurious journey, but makes one constantly aware of the greatness of Chinese food culture traditions. As for foreign tourists in China, regardless of having grand banquets in restaurants or a taste of street-side snacks, they are all direct and pleasurable way of experiencing China.

The True Pleasure of Drinking One's Fill

The Art of Tea

The Chinese have a common saying, "Seven things in the house: firewood, rice, oil, salt, soy sauce, vinegar and tea." It shows that "tea" has already blended into the Chinese's daily social lives and has become a daily consumer's item. Tea contains many vitamins, theine, fluorides and more. Tea can improve vision, clear the mind, and benefit diuretic functions and so on. Not only do the Chinese believe that having tea regularly can prolong life and benefit health, modern science has also proven that tea is a natural health drink good for the body.

The southwest part of China, the sub-tropical mountainous area, is the original birthplace of wild tea trees. At first, tea was only used as ceremonial offerings or food. In the Tang Dynasty when Buddhism was at its peak in popularity and influence, Buddhists discovered that tea can relieve drowsiness when meditating, and can help with digestion. Therefore, tea drinking was promoted, and suddenly, almost every monastery or temple had tea. Soon, tea was accepted by the masses, from highly noble imperial families to the lowly merchants and labor workers, all drank tea. So it goes that "from days of old, famous temples produce famous tea." This is because most temples and monasteries have their own land and agricultural produce. The local believers in the faith help to plant and grow the crops. Improvements in quality of the tealeaves and the promotion of the art of tea-drinking were helped by the rather highly educated and cultured monks. From then on, Chinese tea art along with Buddhism was brought to Japan. Starting in the Naryo period (around early 15th century), Japanese tea art took shape. Korea, as well as countries in Southeast Asia, was also influenced by tea-drinking customs.

Due to differences in culture and geography, the word "tea" in Chinese has mainly two ways of pronunciation. One is based on the northern dialects, which pronounces it as "cha," the other is the southern dialect from areas like Fujian that reads it as "tee." Those countries that imported tea from northern China, such as Japan and India, have their words for "tea" being similar to "cha;" the Russian pronunciation is "chai" and the Turkish say "chay." And those countries who imported tea from the southern coast of China, such as England, has the word "tea;" Spanish word is "té;" the French pronounces it as "thé," and the Germans say "Tee." They are mostly all phonetic translations of the Chinese word for "tea."

Popularization of tea goes hand in hand with its development in trading. In Europe, the earliest tea drinkers were the British. Record shows that at the beginning of the 17th century, the English have already tasted tea from China, and thus aroused a tremendous interest and need in tea supply. In order to secure a continuous supply of tea drinks, the British government ordered its East India Company to guarantee tea supply in stock. As tea gained popularity even among the general public, European countries' need for tea supply increased exponentially as time went by. By early 19th century, China had a tea export volume of 40 million tons to England and created for large Sino-British trade deficits. To reverse the unfavorable situation, British merchants tried every means to obtain opium from India to be exported to China, in exchange for Chinese tea without having to spend real cash. This, in time, sparked the infamous "Opium Wars" which strongly impacted contemporary Chinese history.

Pu'er tealeaves on the tea tree. (Photo by Yunhua Xu, provided by image library of Hong Kong *China Tourism*)

China has, including Taiwan, sixteen provinces and regions that produce tea; including the northern provinces of Shandong, Shaanxi and Henan. Since the Tang Dynasty, northern and northwestern nomadic nationalities began trading horses for tea with tea-rich

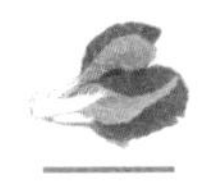

Drinks in Ancient Times
In ancient times the Chinese divided drinks into three categories: thick liquids, alcohol and tea. As early as the Tang Dynasty clear distinctions were made between the uses to which these types were put in normal life. The first was used to quench the thirst; alcohol was used to allay one's cares, and tea was used to clear the mind and raise the spirits. At the same time people became progressively more discriminating about the ways in which tea could be used as an everyday drink. It had healing properties, it could be used in cooking, it could take the place of alcohol, and even in those early days it began to see service as a means of sobering up.

regions, thus spurred a whole new kind of business. It was not until the mid-Qing Dynasty did the horse-tea commerce begin to be replaced by hard currency. At present, tea has become a daily necessity for people in these areas.

Tea is made from leaves plucked from tea trees. Due to different processing procedures, it results in green, red (black), oolong, white, yellow and dark black teas. Famous high-grade teas are products made possible by many factors, including a superior natural growing environment, superior breed of tea trees, meticulous picking methods and exquisite treatment process. While holding high prestige, these high-grade teas also possess important position in the commercial market.

The crucial link in making tea is in its "fermentation." Non-fermented teas are called "green tea," which use fresh tea buds as basic ingredient. Through dry heating or steaming, fermentation of the tea is stopped. Then it is twiddled into strings and dried for the final touch. Tea drink made from green tea is verdurous in color or has a touch of yellow within. Its fresh fragrance is highlighted by a touch of bitterness. Among all kinds of tea, green tea has the longest history with the largest production volume and broadest production area. Of which, the provinces of Zhejiang, Anhui and Jiangxi have the largest production and best quality. Green teas, since ancient times, include a whole line-up of famous high-quality tea including West Lake Longjing, Longting Lake Biluochun, Mount Huang Maofeng, Mengding Ganlu, Mount Lu Yunwu Tea, Xinyang Maojian, Liuan Guapian, all of which are well-known both near and far.

A stream of hot water pours down from high up, brewing the tea and cooling itself in the process. (Photo provided by Rui Huang)

Through fermentation, tealeaves will gradually turn from its original dark green color to a dark reddish tint (black); the longer the fermentation period, the redder the color. Its aroma would also change, depending on the degree of fermentation, to flower fragrance, ripe fruit fragrance or malt sugar fragrance. Completely fermented tea is called "red tea (black tea in the English language)." It is made first by picking fresh tealeaves from the trees, then through a process called "withering," the tealeaves are placed outdoors under intense sunlight. Then it is moved back indoors for cooling, thus the tea becomes much more fragrant. Then the tea is kneaded, fermented, dried and so on through a series of processing stages. Due to its dark red color both when dry and when made into a drink, we call it "red (black) tea." Red (black) tea in its treatment process has gone through chemical reactions, where the chemical contents of the fresh leaves change greatly. Its fragrance is much more apparent than the fresh leaves. More famous black teas include Qimen Red (black) Tea, Ninghong Gongfu Tea, Fujian Minhong and so on. Semi-fermented tea, or oolong tea, are a unique Chinese specialty. The most representative production area of oolong tea is Anxi, Fujian Province. Oolong tea can be distinguished into three grades, which are light, medium and heavy fermentation. Lightly fermented oolong tea has the characteristics of strong aroma, high refinement, making for a golden colored drink. Medium fermented oolong includes Tie Guanyin (literally the Iron Bodhisattva), Shuixian, and Dongding. Its drink color is brown and is steady in taste, giving a rather strong touch to the throat. The heavy fermented oolong such as Baihao Oolong makes for an orange-colored drink with the sweetness and fragrance of ripe fruits.

People in northwest China like the unique taste of "brick tea," which can be enjoyed when dining at a *qingzhen* (Hui Muslim) restaurant. (Photo by Zemin Li, provided by image library of Hong Kong *China Tourism*)

In general, northern Chinese prefer the strong and fragrant huacha (flower tea), or red (black) tea; the people south of the Yangtze River cannot live without Longjing, Maojian or Biluochun; people in the southwestern provinces are used to the pure and rich Pu'er; and people of Fujian and Guangdong like to use oolong to make "gongfu" tea. The nomadic people of China drink milk teas of various types. Some people say that the green tea symbolizes the strong scholarly air of the Jiangnan (south of the Yangtze River) people. Red (black) tea, on the other hand, has a feminine quality, bringing a sense of peace. The oolong symbolizes the wisdom of the old, rich and polished. Drinking huacha feels rather frolic as if walking the busy folk streets, with a direct and full-bodied taste. Therefore, perceiving a Chinese with his or her tea can give many clues about where the person is from, the individual's personality and level of self-cultivation.

In most tea-producing regions of China, the growing of tea trees and the plucking of tealeaves have its seasons. The plucking of tea usually takes place in spring, summer and autumn. Tealeaves from different seasons have different appearances and inner quality. Tealeaves plucked in spring, from early March to the Qingming Festival (also known as Pure and Bright Festival or Tomb Sweeping Day, around April 5th of each year), are called "pre-ming tea" or "first tea." Its color is of light jade green, and tastes pure with a touch of acerbity. Two weeks after Qingming, it is the Guyu solar term on the Chinese lunar calendar. During this time, the Jiangnan area will experience a round of fine precipitation for the moistening of crops. And this brings forth the second peak season of tea picking. Tealeaves collected after the Qingming but before Guyu are called "pre-rain tea," and the spring tea picked after that are called "post-rain tea." Spring tea's prices usually vary according to the time the tealeaves were picked, with the prices being higher for earlier tea and lower for the later. In most cases, early-spring green tea is the best in quality among all available tea throughout the year. Tea from the same year is considered new, while tea of more than a year old is considered aged. Green tea and oolong tea are better when fresh, while aged Pu'er tea also has great flavor. Tea lovers have different choices throughout the seasons. Green teas are great for spring, while Chrysanthemum tea (chrysanthemum flower dried under direct sunlight for soaking into drinks, Hangzhou in Zhejiang Province produces the most famous chrysanthemum tea) accommodates nicely for autumn. In late fall and winter, oolong, Pu'er and Tie Guanyin are great for the cold weather. Throughout the year, there are more varieties of tea than one can choose from. Experienced tea drinkers not only can discern between new and aged tea. They are even able to tell during which the season the tea was picked from.

A young girl of the Bai nationality boiling tea. (Photo by Guanghui Xie, provided by image library of Hong Kong *China Tourism*)

Tea drinking is deeply rooted in the Chinese nature. In mid-Tang Dynasty, a scholar by the name of Lu Yu (733–804 AD), who had spent his childhood in a monastery, collected and compiled older writings concerning tea and combined it with his own thorough studies. Eventually he completed the first work of writing in the world relating to tea, titled Cha Jing. This classic work systematically recorded the properties of tea trees; the traits of tealeaves; tea-growing, picking and making; tea-brewing techniques and tea drink ware; experiences with tea and so on. The manuscript also introduced the origins of tea as well as tea-related matters before the Tang Dynasty, labeling tea production areas at the time. The book is an important work of contribution to Chinese tea culture.

In the middle of the Tang Dynasty, there also emerged a competitive way to judge the quality of tea and tea-brewing techniques, as it was called ming zhan, literally meaning "tea war." It was a reflection of the highest form of tea sampling in ancient times. In the Song Dynasty that followed, when tea drinking was the prevailing trend, it was a time in history that people paid the most attention to "tea battles." From kings, generals and prime ministers to the common folks, all participated in the art. Not only were there competitions between famed tea production areas and respected temples, even at street markets selling tea, people would have to "duke" it out, and it was closely connected to the trade. Many kinds of famous tea and tea used as imperial tribute came about either as a direct or indirect result of the tea battles. In a tea battle, usually two to three people would gather together, each presenting one's best available tea. After heating water and brewing, the best tea makes the winner. Tea art emphasizes many ideas, as "freshness is noble" in tea stock and "liveliness is noble" in the water used. The taste of tea depends on "fragrance and smoothness" for the best quality, and the fragrance of tea should be the "real fragrance" of the tealeaves. As for the color of the drink, "pure white (clear)" is the superior kind. Correspondingly, using black porcelain teacups (Fujian-made Black Porcelain from Jian Kilns) became the leading fashion that replaced the formerly used blue porcelain. As a result, the value of a teacup was not only in its aesthetic appearance, more importantly was its power to produce an unforgettable experience at times of enjoying tea, directed by the sense of touch, sight, smell and taste. When making tea with a black porcelain teacup, the beauty of a white liquid casts a great sense of aesthetic satisfaction. It is an art appreciated by all social classes, from imperial tea feasts to gatherings of merchants and servants. This trend even found its way into Japan. Tea battle has had a profound impact on the development of Chinese tea culture.

The standards for sampling, evaluation and inspection of tea in many ways were derived from the book of Cha Jing and tea battles. To make a fine pot of tea, not only are high-grade tealeaves to be used, the water quality, temperature, quantity and the type of tea ware must also be taken into account. The ancient Chinese believed that spring water from high up in the mountains is best for making tea. River water, ice melt, and rainwater are second in quality; with water from earth wells being the worst. In modern terms, best-quality water means fresh soft water with low mineral content, while hard water with high mineral content should not be used. Required water temperature should be adjusted accordingly with each kind of tea. For most tea, close to 100 degrees Celsius would be proper. However, for green teas and teas with a low degree of fermentation, water temperature should not exceed 90 degrees Celsius. The amount of tealeaves used to make a drink also depends on the type of tea used. From a quarter to three quarters of the teapot's capacity are all possible. As for "tea ware," different types of tea require different vessels for the best experience. For huacha, porcelain pot is used so as to seal in the fragrance. Green tea is light in taste, and zisha (literally "purple sand") earthenware pots absorb taste and fragrance easily, so it is best to use glass to preserve the fragrance and also allow for a clear view of the tea's color and form in water. As for black tea and semi-fermented tea, the best utensil to use would be clay pots. To really comprehend

Old men in a teahouse in Chengdu. (Photo by Hongjiang Zhang)

the enjoyment of drinking tea, and feel the true taste of tea, requires very high culture and artistic cultivation in an individual. One can earn from the process certain artistic enjoyment to eventually find self-cultivation and enlightenment. Therefore, drinking tea is the embodiment of the Chinese view on life as an art.

When it comes to tea ware, before the Tang Dynasty, tea and food vessels were indistinguishable. As tea drinking-grew more popular, tea containers became more and more refined. By the end of the Tang, the most ideal tea ware was invented, the zisha (purple sand) pot. It is different from most earthenware as it uses extremely fine maroon-colored clay as raw base material. By skillful craftsmanship, it becomes a brownish-purple pot with a fine and smooth touch, and gives a primitive and simplistic aura of elegance. This kind of pottery, made by heating to about 1,100 degrees Celsius, has no glaze on either the inside or the outside. Looking at it through a 600 times magnification microscope, small pores can be observed on its surface that allow for the passage of air but not water, thus keeping the fragrance of tea sealed inside. With some literati and refined scholars at the time directly participating in the design and making of the pots, these pottery pieces combined poetry and rhymes, paintings, seal impressions and sculpture into one, possessing very high artistic and functional value.

The reason that the porous zisha pots became famed throughout the land after the Ming Dynasty has to do with changes in tea-drinking practices. At the time, drinking tuancha, or bunched tea in a discus shape, was giving way to drinking loose tea. But using small cups to brew loose tea was unsanitary and difficult to maintain temperature, so teapots were used. Using small teapots to make tea is a tradition that has started in the sixteenth century and continued to the present day, with already over four hundred years of history. When using zisha pot to brew tea, its low conductivity of heat and a whole on the lid prevents drops of water vapor from coagulating under the lid and ruining the taste when dropped into the tea. Since the teapots have already been heat-processed when being made, it would not crack or break even when it is heated on a stove. The longer a zisha pot is used, the more brilliant and smooth it becomes; tea made from it will have a stronger fragrance. Teapot aficionados like to use different pots for making different teas, so as to keep the seasoning of the teapot pure and consistent for a long time.

The birthplace of zisha pots is in the famous "pottery city"—Yixing. It is situated at the common boundary of Jiangsu, Zhejiang and Anhui provinces, by the shore of the Taihu Lake. In the Tang Dynasty, this place was already a famous tea production base, from which many famous types of tea were offered to the imperial courts. Yixing's zisha pots became widely known in the Northern Song Dynasty. By the Ming, there were many master pot-makers in Yixing, producing many pots with exotic forms and simplistic elegance that were the benchmark for tea ware. Most tea lovers usually like to collect and enjoy teapots. Certain exquisite teapots by the hands of famous makers can be priced as high as pure gold. Collecting pots or "raising" pots is perceived as an elegant hobby until this very day.

What is worth mentioning is that in the Minnan and Chaozhou areas in southern China where people make gongfu tea, zisha pots from Yixing have been very popular ever since the Ming and Qing dynasties. There, finely crafted upscale zisha pots were once the symbol of education, status and social role for local men. Regardless of high officials, the wealthy or common citizens, they all leave no stones unturned in getting their hands on a zisha pot and treasure it as a jewel of dear value. Some even take it with them to their graves. However, in Jiangsu, the production area of the pot, people prefer green tea. As techniques of tea-making improve, there are very few people today who still use zisha pots for making green tea. Rather, they prefer white porcelain cups or glass. Zisha pots are now mostly regarded as an object of art to be appreciated in one's home. For a top-quality zisha pot, people would rather pass it down the generations within the family, seldom would anyone keep it as a personal funerary object.

China always had the custom of "treating arriving guests with tea." Some also advocates having tea in place of alcohol. The way of offering tea is really very simple. Before making tea, ask for the preference of the guests. The water used to make tea should not be too hot so as to burn the guest. When pouring tea, the rule of "full cup for wine and half cup for tea" should be observed; leaving one fifth of the cup's capacity unfilled should do the trick. When the host pours tea for the guest, the guest uses the index and middle fingers to lightly knock on the table to show gratitude. It is said that this custom was passed down from the Qing Dynasty and is not only popular in China, but also popular among overseas Chinese of Southeast Asia.

Gongfu tea is a unique custom of the Chaozhou region of Guangdong Province. It has existed since the Tang Dynasty. It is not just the first line of courtesy, as people of Chaozhou who travel or reside overseas use it as a ways of paying respect to the ancestors. Authentic Chaozhou gongfu tea sincerely abides by the old traditions, usually limiting the number of participants including host and guests to four. This is congruent with the older ideas of tea enthusiasts of the Ming and Qing dynasties, where "harmonious thoughts in different hearts" is advocated. One should not think too many ideas. When guests take seats, it must be in the order from senior to junior or from high to low status, starting from the right side of the host and make two rows. After the guests are seated, the host starts to work his magic. Not only is tea ware as fascinating as appreciating antique, the quality of tealeaves, the water and the brewing, pouring and drinking of tea all very interesting areas of study. The teapot used in gongfu tea is small and exquisite, only about as big as a fist. The teacups are even smaller at about half a Ping-Pong ball size. The kind of tea chosen is the oolong, which is complete in terms of color, fragrance and taste. Tealeaves are stuffed tightly into the pot, almost filling it all. It is said that the tighter you push the tealeaves together, the

A Sichuan teahouse full of life. (Photo by Jin Chen, provided by image library of Hong Kong *China Tourism*)

stronger the taste will be. It is best to use water that has been settled to make tea. When making tea, immediately pour boiled water into the pot. The first couple of rounds of tea water are not drinkable as it is for rinsing the tealeaves and the cups. When pouring tea, do not finish pouring into one cup and go on to the next. Rather, one should alternate between the four cups in order, filling each little by little until all cups are about seven-tenths full. When the thickest tea water is left, containing the essence of the tea, one should evenly distribute it into each of the four cups to assure equally strong taste and uniformed fragrance. There exists a rule when drinking gongfu tea. One should not drink the tea immediately, but should rinse one's mouth with cool water to guarantee tasting the true flavor of the tea. When drinking, one slowly sips and uses the tongue to feel and taste the tea. Gongfu tea is very strong and contains strong base, therefore one would feel bitter and astringent at first. But as one drinks more, the tea becomes more fragrant and rather smooth and sweet, as one starts to feel more energized. While having gongfu tea, people can chat to their hearts' content, and should feel peace of mind. This is the true meaning of gongfu, and also demonstrates tea art's uniquely Chinese characteristic of upholding nature and freedom, reflecting a special Chinese kindness that is honest and rich, with long-lasting charm.

From gongfu tea, one can easily connect in mind various types of teahouses. In China, running a teahouse is a very popular service profession. Especially in the Jiangnan area, there are teahouses to be found in every corner of every small village, town and large city. There are teahouses that have kept to the centuries-old traditions, others combined features of cafes and bars. And a rather large percentage of teahouses provide dining services. When looking back in history, teahouses flourished beginning in the Song Dynasty. At that time, there were teahouses for every social class. Upscale places not only have paintings and calligraphy pieces adorning its walls, but also all kinds of fresh flowers, bonsai trees inside, plus background music for an elegant feel. Upon reaching the Qing Dynasty under the reign of emperors Qianlong (1736—1795 AD) and Jiaqing (1796—1820), Beijing's teahouses combined music and folk arts for a real treat. Customers can drink tea while enjoying live entertainment, or can bring their own tea and pay for only water. Therefore, many opera theatres in Beijing were once called "Tea Gardens." Beijing's specialty, the Dawan Cha (Large Bowl Tea), is now hard to find. But in the old days, one can have a bowl of Dawan Cha under just about any tree while sitting at a shabby table on a shabby seat with a rough large bowl in hand. People of Sichuan have a long history of tea-drinking, and teahouses are very popular. In the provincial capital, Chengdu, teahouses range from the smallest with three to five tables to the largest with several hundred seats. There, people use gaiwan, a complete set with tea bowl, stand and lid. Using long-mouthed bronze pots to pour tea is an absolute Sichuan specialty; a long jet of tea water enters the bowl and stops just as it fills to the mouth, with not a single drop wasted. Elderly men prefer enjoying tea while watching Chinese opera performances or chatting with friends. Modern white-collared professionals like teahouses for it gives them a chance to relax, socialize or talk business. There is a common saying in China, "Tea cleanses the heart." The peaceful and quiet state of mind that tea represents is different from mundane society, full of blatancy and flippancy. People who like tea can easily find a state of purity and contentment.

Wine, the Beverage of Romance

Alcohol drinks are a kind of material culture shared by all nationalities of the world. Before the advent of distillation machines, wine could only be made the primitive way. Using crops to make wine is a special characteristic in Chinese alcohol-making history. The yellow wine, or rice wine, being one of the three main kinds of alcohol beverage (rice wine, grape wine and beer), is known as the model of oriental winemaking.

Winemaking and drinking originated in China long ago. Ancient writings point to multiple origins of alcoholic drinks, but only a small portion can be taken as true history. In common society, Dukang is worshipped as the god of wine, as he was the one who first made wine. However, as early as the Shang Dynasty, the Chinese already widely practiced the making of alcohol. From existing oracle bone and bronze inscriptions, we learn that many people of the Shang used wine as offerings to ancestors. At the same time, drinking wine was already popular. In more recent archeological excavations, Shang Dynasty winemaking sites were discovered. In 1980, in Henan Province, ancient wines from the late Shang period (about 3000 years ago) was discovered in archaic tombs, and are now kept at Beijing Palace Museum. It can be considered the oldest wine in China. Being large in land mass and with abundant natural resources, China's different agricultural crops, water quality and winemaking techniques in each region, gave birth to many excellent types of liquor throughout the land.

A very important winemaking innovation by the ancient

Chinese was the use of yeast. The primitive forms of yeast were molded or germinated crops, mainly wheat and rice. People reformed the molded crops to make distiller's yeast. The yeast contains bacteria that turn starch into sugar and saccharomycete, which facilitate the forming of alcohol. Different kinds of yeast are adopted in different regions, making for more varieties in wines. In the Northern and Southern Dynasties (420—589 AD), winemaking techniques have already reached tremendously high levels. The book Qimin Yaoshu recorded a dozen ways of making wine yeast.

This natural way of wine fermentation, after thousands of years, has become quite a proven technique. Its basic winemaking principle and technique is still in use today. Using this method to make wine relies mostly on experience and is limited to small-scale production, as it is usually performed through manual labor. Wines produced in such a way have no exact scientific testing standards.

The basic ingredient for rice wine varies by region. In the north, it is sorghum, millet and glutinous millet. While in the south, mostly rice (sticky rice being the best choice) is used. The wine's alcohol content is usually around 15 proof, and becomes tastier as time goes by. Yellow wine's color is not always yellow, as some are black or red. When filtering technique for wine has not been fully developed, wines are usually muddy in transparency. The ancients referred to such as "white wine" or "turbid wine."

Starting in the Song Dynasty, Chinese culture and economic centers moved southward, the production of yellow wine became even more prevalent in the southern provinces. By the Yuan Dynasty, spirit drinks popularized in the north as yellow wine production slumped. The southerners do not drink spirit as much as the northerners, therefore yellow wine production remained high in the south.

Shaoxing Yellow Wine being transported to other regions by way of water channels (Photo by Guanghui Xie, provided by image library of Hong Kong *China Tourism*)

In the Qing Dynasty, yellow wine made in Shaoxing of Zhejiang Province dominated the domestic and even overseas markets. Even now, drinkers of yellow wine still prefer "Shaoxing Yellow Wine."

In many parts of China exist families that have the tendency of making their own wines. It demonstrates just how popular the method of using yeast in winemaking is. Some devoted wine drinkers believe, truly delicious wines come not from wineries, but from hands of skillful common people. Adding "major yeast (made with wheat, barley etc.)" to long-grained Indica rice inside a wine jar and sealing it for over a month, rice wine of about 40—50 proof can be made. Using "minor yeast (made with rice)" with glutinous rice and sealing it for several days would produce fermented glutinous rice wine of about 10 proof; if sealed for over a month, it would yield sweet wine. Regardless of rice wine or sweet wine, the longer it is sealed, the stronger the flavor. Fermented glutinous rice wine is simple to make and is an inexpensive and delicious tonic drink. The drinking of fermented glutinous rice wine in southern China is quite popular. As always, many try to look from a medical perspective, believing in alcohol's healing ability. Medicinal wines are made to improve the circulatory system and preserve health.

Traditional Chinese white liquor (spirits) is the most characteristic of distilled alcohol. At around 6th to 8th century, China already has its distilled liquor. Primitive distillation was another one of the Chinese contributions to winemaking. From the end of the 19th to early 20th century, after the introduction of microbiology, biochemistry and engineering sciences from the West, China's traditional winemaking technology experienced great changes. Mechanization standards greatly improved and production yield increased as a result. Guizhou and Sichuan provinces in southwestern China are the two publicly acclaimed provinces in production of superior grade white wine. Even so, due to differences in natural produce, every region in the north and south use different base ingredients. Nearly every province would produce its own unique label with tastes to match the local provincial inhabitants. Hence Chinese wines come in no less than

Many supermarkets supply imported grape wine. (Photo by Weijian Lin, provided by Imagechina)

forty to fifty varieties, much more than what is known internationally, namely the Luzhou Laojiao, Guizhou Maotai, Shangxi Fenjiu and Shaanxi Xifengjiu. China's oldest beer brewery was built in 1900 in the city of Harbin. Even though beer brewing has only been in China for less than one hundred years, beer is already the top selling alcohol in China.

Tsingtao (Qingdao) Beer is an internationally recognized Chinese beer brand. The picture shows a street-side advertisement for Tsingtao Beer at Tongluowan (Gong Bay) in Hong Kong. (Photo by Lei Xin, provided by Imagechina)

Wine since ancient times were closely interwoven with peoples' daily lives. People use wine to pay tribute to ancestors, showing respect; or for self-enjoyment while writing poems and composing rhymes; or when treating friends and family, heightening the lively atmosphere. Wine undoubtedly holds a very important place in Chinese culture and living

Ancient kings and princes' banquets and feasts cannot go on without wine. All kinds of wine vessels therefore became important objects of courtesy. The most important of which were the bronze *Jue*, *Zun*, Yi and other drink containers which symbolize social status and ranks. From archeological findings around China, bronze wine vessels were once the fashion of the times. The lifting of the ban against wine for the common citizens usually takes place during times of changes in dynastic rule or the succession of sovereignty, as well as important imperial activities.

A beer store on the street. (Photo by Guoqiang Ji, provided by Imagechina)

The Chinese in ancient times often used crops to make wine. Therefore a good or bad harvest during the year strongly influenced the decisions of the dynastic rulers to either impose wine ban or certain amount of wine tax. And correspondingly, a strong or weak year for wine production reflected the crop yield of the year. Wine in the past was directly related to people's livelihood and taxation. Since the third year of the Tianhan period under Emperor Han

Distilled rice wine from sorghum is rather common. (Photo by Baoxiu Shi, provided by image library of Hong Kong *China Tourism*)

Wudi (98 AD), after the central court exercised the exclusive right to sell and buy wine, duties collected from the wine industry became an important source of government treasury revenue of subsequent feudal dynasties.

Wine and most Chinese literati have always had an intimate connection. In medieval times, there were many written accounts of the important personage of the Wei & Jin Period (220—420 AD) and poets of the Tang Dynasty being wine lovers. It was the two very important periods linking "wine and Chinese culture" together in history. In actuality, the connection between the literati and wine did not begin in the Wei & Jin Period. However, it is still rare to find binge drinkers such as the "Seven Wise Men of the Bamboo Forest" who drank heavily for no special reason. Drinking took up much of the time of these men in the Wei & Jin Period. Those were people who lived during socially turbulent times, who used wine to ease their worries for society and escape from misfortune. Sometimes they expressed zealous objections about the government after having drinks. Such acts are a reflection of the helplessness of the literati class during those restless years. From then on, binge drinking by scholars is no longer viewed as corrupt and hideous, rather regarded as admirable and romantic. It seems as though all Tang Dynasty poets enjoy drinking boldly. Famous poets such as Li Bai and Du Fu are all great men of liquor known throughout China and abroad. Their poems have the nature of wine and wine fuels their creativity. Traditional Chinese art forms such as poetry, music, painting, calligraphy and others are all very emotionally expressive. Wine can help the artist achieve the uninhibited natural and honest state of being, stir up their creative juices. So people now imagine a very romantic connection between wine, poetry and the literati.

The Chinese pay attention to the "drinking mood," something that one must have to truly enjoy drinking as a real charm of life. "One thousand cups of wine is too few when with a bosom friend," a common saying that suitably embodies the emphasis which the Chinese places on harmony of relations between people, meaning to share with others moments of joy. Wine enriched the affections of the Chinese. Playing finger-guessing games, composing impromptu music, poetry or even dance, are all fun-adding games for drinking during banquets. They are also the highlight of Chinese drinking habits. Both sides playing the drinking games, sometimes, are like two opposing armies, swinging their arms and jutting their fists, while howling the game lyrics. The game is a match of wits, courage and alcohol tolerance, and is truly quite fun. Dining together and playing drinking games has become a unique and favorite pastime of the Chinese, its objective is to communicate friendship and increase family love. So a banquet could last for quite a long time, ranging from a couple of hours to the entire night.

Hospitality of the Chinese is expressed to the fullest extent at a banquet with alcohol drinks. The communication of affection most often gets a boost of sincerity and directness. Many places welcome and treat guests with wine. When old friends reunite and when friends meet, a few cups of wine can be most delightful as wine produces an air of warmth and harmony. "Bottom's up" is a custom practiced widely in both south and north China. When a banquet begins, the host usually delivers a few words of welcome, follows by the first toast. The host first finishes his cup until the last drop, in what we call "finishing first as respect" for the guests. Sometimes, the host will also need to propose toasts to the guests individually in the order of importance. Anyone not returning the favor would be considered disrespectful and would often be subject to punishment in the form of more drinks. Therefore, guests must return the toast to the host. Guests can also propose toasts amongst themselves. Moreover, it is best not to be tardy for a banquet, or the host and guests will suggest punishing the late attendee with many drinks. When proposing a toast, the initiator and the receiver must all stand up. Most toasts are limited to three cups. The more the guest drinks, the happier the host will be. A very interesting thing is that the toast initiator would like others to drink more than he or she. Especially for some very hospitable minority nationalities, drinking unrestrained is a "must do" when with a guest. Take the Mongolian nationality for instance, the host often hold bowls of wine in both hands while singing the toast song, and keep feeding drinks to the guests one by one until they are all completely drunk. The Miao, Dai, and Yi nationalities of southwestern China practice a "sucking" method when having alcohol. This involves using a long reed stalk or bamboo shoot to suck from large wine jars or pots. It is usually done in the order from the eldest to the youngest person. Wine holds another clever use among the minority nationalities in one of the oldest traditions. When becoming blood siblings or swearing oath of alliance, chicken or sheep is slaughtered, or sometimes arms of the oath takers are cut, to allow blood to drip into bowls of wine. The minorities see blood wine as a sacred bonding element for people who drink it.

People who can still keep their composure and their gentlemen or ladies' charm, under the influence of alcohol, would be deeply respected. Confucian thinking emphasizes the "wine virtues" for which the drinker is to uphold. Confucianists do not oppose having alcohol; using wine to pay tribute to ancestors, to provide for the aged and pay respect to guests are all considered virtuous acts. But to save crop supply, one should constrain the amount of wine used. Being overly drunk and unable to tell real life from alcohol-produced illusions is not an attitude favored by the Confucianists, who abides by the strict rule of "wine to deal respect, to treat illnesses, and to bring joy." On special occasions, wine is indispensable. However, it is viewed as an item of luxury, since without it, daily life would not be impacted. There also exists

A bar street by Zhujiang River in Guangzhou. (Photo by Guosheng Zhang, provided by Imagechina)

a popular belief that "wine can disrupt one's nature." Since wine is addictive, large amounts of consumption can cause inebriety, which often lead to stirring of trouble or harm to health. People therefore see it as the source of disorder. So from ancient times to the present day, there has never been a shortage of people who advocates drinking morals and manners, and who conducts alcohol education and advise against excessive drinking. In current times, some government agencies have clearly imposed restrictions against their civil servants from drinking during lunch hour on a workday. For other specialized professions, even more definite alcohol restrictions exist. Drivers who drink and drive will be prosecuted by law.

Chinese drinking courtesies and customs were born almost at the same time as wine was invented. Some customs have been kept until today. "Marriage wine feasts" have long been synonymous with weddings. To prepare for "marriage wine feasts" is the same as preparing for weddings. To drink "marriage wine" means going to attend a wedding. At a wedding banquet, the bride must propose toasts to the parents and guests. The newly weds must also have "arm-crossed wine" to imply "a hundred years of" happy marriage. On the third day after the wedding, the bride must take the groom back to her parents' home. The bride's family will host a banquet to welcome the guests; this is called "homecoming wine." For a newborn baby, "month-old wine" or "hundredth-day wine" is popular banquets held for celebration according to Chinese tradition. When the baby is a month or one hundred days old, the parents of the child will put on a few tables of feasts for treating family and good friends. Most guests will bring gifts or will wrap money inside a small red paper envelope called a "red bag," for the child's family. "Longevity wine" is a birthday feast prepared for elders in the family. A senior of sixty, seventy, eighty, ninety or even a hundred years old can be called "da shou," or "grand longevity." Most times, the banquet is prepared by the elder's sons or daughters, or the grandchildren; attendees include family members and dear friends.

Each of the several major holiday celebrations of the Chinese all has its corresponding wine feast and celebration. On Chinese New Year's Eve, people drink "New Year's Wine," wishing for good health and closeness in the family in the New Year. On the fifty day of the fifth lunar month, the Dragon Boat Festival, people will have changpu wine (wine made from Gladiolus hybridus, an aquatic plant from which fragrant oils can be extracted. Changpu wine is a compound drink made by taking changpu fluid as flavoring, or directly mixing it with yeast made from barley and pea, to make sorghum wine after immersion and soaking) to ward off evil and bring peace and security. For Mid-autumn Festival on the fifteenth day of the eighth lunar month, whether it is uniting with family or meeting with dear friends, drinking while admiring the full moon is to be a part of the evening. This is also the time when sweet-scented osmanthus flowers are in full bloom. So drinking osmanthus wine is also a part of the Mid-autumn tradition. On the Double-Ninth Festival, the ninth day of the ninth lunar month, there has been the custom to climb to great heights and enjoy wine; many regions prefer having "chrysanthemum wine" on this day.

Westerners pay attention to having different kinds of alcohol on different occasions. Alcohol is sometimes the symbol of social status in the West. However, this practice is not suitable for China. Even though Chinese liquors have different grades from low to high quality, Chinese people select liquor by personal preference for the fragrance, taste and texture, and will seldom use the year of production, color and place of production as a basis of judgement.

Alcohol drinks not only fused with the daily lives of the Chinese, people even use it to communicate different feelings and thoughts. It can be said that wine is all-encompassing in reflecting actions and emotions of humans. The feelings, either sorrow and pain, or the joyous wonders, all reserved only for the drinker to experience.

The Chinese's wine culture is of long-standing history and well established. Wine influenced the ways in which the Chinese live, and have shaped the personality of the Chinese. Particularly in recent decades, with accelerated developments in China's economy, people's lifestyles are becoming ever multi-faceted. With traditional winemaking techniques and drinking habits still earning the favors of the public, imported wines, beers and other types of liquor from foreign countries are also gaining popularity. When friends and family gather to enjoy a good drink, the selection of alcohol drinks has imperceptibly broadened. Thus drinking becomes more enjoyable and making China's liquor-culture even more vivid. The bars and pubs springing forth around the country in recent years represent the spending preference of the younger generations. Many foreigners just setting foot in China are awed by the popularity of bars in China's major cities. The bars' internationalized styles also reflect the Chinese's carefree and open lifestyle.

New Dining Trends

In recent years, with the prospering economy and the improvement of living conditions, lots of Chinese people have changed their requirements and methods of diet, laying more emphasis on the balance of food and health in their daily diet. An acuter awareness of health not only influences people's dining habits and diet structure, but also boosts the development of green and environmental-friendly agriculture, and accelerates the adjustment of the relevant food industries. On the other hand, as a part of the western consumerism culture, the western catering industry keeps expanding in the Chinese market. More and more Chinese people have got the opportunity to taste their affordable western delicacies. The western-style fast food has also found its way into China, a nation renowned for its typical, oriental style of "slow food". All of these are changing the Chinese way of life, and the new dining trends are accelerating the continuous innovation of Chinese catering industry.

Workers eating in cafeterias in the 1950's. (Photo taken in 1958, provided by Xinhua News Agency photo department)

The many restaurants all over China, each has its own uniqueness. In general, a mutual trait shared by all is that the entire food service industry is facing the challenges posed by the change of our times and new ways of living.

Around the 50's and 60's of the 20th century, all industries in China, including the restaurants have adopted the policy of joint ownership and management by both the state and private entities. To keep consistency in cooking standards, restaurants that used to keep secrets from one another began to disclose information and communicate for improvements. Some lost skills and forgotten dishes were rediscovered, welcoming a new golden age of food culture. However, due to certain social attitudes and trends at the time, opposing luxury and encouraging austerity, those that paid attention to food were labeled as corrupt and backward thinking. The urge to enjoy a tasty meal at restaurants was being suppressed, and culinary art development unavoidably suffered. Most state-operated hotels and restaurants abode by traditional convention, served mostly authentic styles with limited choices and high prices. However, customer services at most state-owned dining places were most often unsatisfactory.

By September 30th, 1980, the first privately held restaurant since China's reform was opened in Beijing, named the "Yibin Restaurant." At that time, all other restaurants were state-owned. Cereals, oils and tofu required government-issued food certificates and coupons. So the new private restaurant stirred quite a reaction from the international community. Even to this very day, the owner remembers down to the last details of the day his restaurant opened. He spent 36 Yuan, bought 4 ducks, and made several duck dishes. Only a few days later, customers of the restaurants included ambassadors from 72 countries and press correspondents of 74 different media.

As living conditions improved, people still wanted to dine out for a change of taste, and have some foods that one does not know how to or does not have the means of making at home. Restaurants

During the planned-economy period in China, the government issued "food notes" starting in 1955, in order to guarantee the supply of foodstuffs for citizens in cities and towns. Along with Chinese economic development, prices of grains and oils were no longer tightly controlled. By the 1990's, "food notes," which existed then in name only, were altogether abolished.

of different tastes and different grades operate for survival, as more and more newcomer rush to the scene from different parts of the country. The food service industry became China's hotspot of investment. Investors and managers sought help, from chefs of the older generations or epicures of famous schools and families, for the ways of making traditional dishes. Restaurants that used to offer handmade dumpling or noodles were expanding their menu offerings, starting to simultaneously sell other types of foods. Attractive hostesses wearing striking uniforms or qipao (a traditional Chinese dress), elegant hats, and a band across the chests marked with the names of the restaurants, greet guests with a friendly smile in front of newly opened restaurants. Since privately held restaurants pay the utmost attention to quality service, waiters and waitresses treat customers with warm, adequate courtesy, so as to attract much business. In contrary, many state-owned restaurants situated at noisy districts gradually lost their competitive edge due to their rude service and plain menus. In no time, the great tasting foods that disappeared from the cities returned once more. Certain restaurants in northern China, serving traditional foods, started to use their "old brand names" again. In Shanghai, food houses use the word "authentic" to attract diners.

Before the 90's of the 20th century, people focused on just the food when dining out. Even for street stands and grills, people would form lines just to have a taste as long as the prices are fair and the servings are big enough. However, as economic development in China's cities and towns flourished, consumers have set their eyes on more than just filling the stomach. Most people, on top of great food, look for elegance of environment, cleanliness and thoughtful service.

Great tasting home-style cooking once again ventured out of the homes and into the market place, onto the menus of restaurants. The once gone tastes were now all back in the cities. Compared to the rare delicacies of high-class restaurants, home-style dishes have no special characteristics in taste, but many restaurants flaunt it as a customer magnet, with great taste and affordability. But the most attractive aspect is the home-like feeling that it produces, making dining in restaurants no different from home, free and cozy. Most often, home-style restaurants are rather small in size. Its simple menus only offer well-known dishes such as *Gongbao* (Kungpao) Chicken, *Yuxiang* Pork, Fruit Salad and so on. In recent years, people's daily diets show obvious changes as feasts for birthdays, gatherings and treating guests are moving to public places. Snacks, home-style dishes and fast foods are quite popular. The rise of home-style cooking not only changed the dietary habits of common citizens, but also injected new life into the highly competitive food service industry. All kinds of signs with the words "Home-style Food" written in all fonts and styles adorn the street sides, with an ever-expanding scope of business. Famous names include "Mao Style Home Cuisines," "Guolin Home-Style Foods" and so on. Home-style gourmet changed from daily meals for the common

Chinese-style fast food. (Photo by Roy Dang, provided by Imagechina)

people into commercial food, restaurant food.

As home-style dishes developed, the ways in which home-style restaurants managed their businesses changed. Some restaurants introduced more expensive foods such as Beijing Roast Duck, while other restaurants kept the familiar "old faces," such as Simmered Pork Ribs in Soy Sauce, Casserole Bean Curd, Green Pepper and Potato Strips and so on. High-priced cuisines not only have differences in basic ingredients from home-style dishes, its ways of making are also much more complicated.

To cater to different tastes of different customers, in cities with large volumes of migrant population, such as Beijing, Shanghai, Guangzhou, Shenzhen and so on, more and more restaurants are offering regional cuisines. Food culture and fashion met with immense changes in every one or two years. First it was the Yue (Cantonese) cuisines, popularized throughout the country; followed by Fish with Pickled Vegetables belonging to the Sichuan style; Lamb Kabobs from Xinjiang; Mao Style Simmered Pork of Xiang (Hunan) cuisine; Henan's Simmered Lamb in Soy Sauce; Mala (Hot and Mouth-numbing) Hotpot from Chongqing; Dumplings from northeastern Regions of China; Shanghai's benbang foods; Hangzhou cuisines; Boiled Fish in Hot Oil from Sichuan; Fragrant and Hot Crab; Yunnan cuisines, Guizhou food, Taiwanese food and much, much more. Changes in popular taste were like gorgeous fashion models on the catwalk, attractive and ever-changing. Since a couple of year ago, places such as Beijing, Shanghai and Taipei welcomed a new kind of home-style cooking called "Personal Home Dish." These restaurants make original dishes and snacks for their guests, while providing a private, all-

Chinese-style fast food. (Photo by Roy Dang, provided by Imagechina)

to-yourself kind of atmosphere with high emotional appeal. Most of these places are small in size; some even adopted membership policy, while others require reservations. These restaurants with a personal touch are frequented and loved by white-collared urban professionals. Combining the practice of Western cocktail parties, buffets let the Chinese experience a meal with the freedom to choose; a great contrast from traditional Chinese restaurants. But the biggest advantage of buffets is that people no longer have to sacrifice their own taste preferences for the sake of others; making social intercourse while indulging oneself only with his or her favorite food a possibility.

For those foreigners new to China, aside from trying tasty Chinese foods, even more enjoyable is to try some "old brand names" of food houses, where there is a strong Chinese cultural experience. These old names include Quanjude, Bianyifang, Donglaishun, Fengzeyuan, Fangshan (imitation imperial cuisines), Liuquanju, Shaguoju, Kaorouji, Kaorouwan, Gongdelin of Beijing; as well as the Old Restaurant of Shanghai, Laozhengxing Restaurant and Meilongzhen, all located in Shanghai; still more are Tianjin's Dog-Won't-Eat Stuffed Buns, Hunqishun Restaurant, Tianyifang Restaurant and more. Most of these food houses have several decades to well over a hundred years of history. Though it has been through the changeover from joint state and individual ownership to privatization, these businesses retained their competitive edge in the food service industry. "Old names" attract customers not only with featured dishes, but rely more on its historical and cultural meanings.

The most famous of the Beijing Old Names—Quanjude Roast Duck Restaurant, is a characteristic example. In actuality, the oldest name in roasting ducks should be Bianyifang Roast Duck Restaurant. Quanjude appeared slightly later but exceeded the former in business. Especially in the eyes of the foreigner, Quanjude is the most acclaimed roast duck house. On basis of preserving the hanging roaster stove, Quanjude Roast Duck Restaurants' business

A restaurant serving authentic French food and Western sweets. (Photo by Shenghui Liu, provided by Imagechina)

multiplied by the day, with over one hundred locations throughout China. It even created its unique all-duck feast. Many people like to have Quanjude Roast Duck not just to have a great taste, but also to savor the century-old stories of the past.

The Quanjude roast duck restaurants, while not abandoning their speciality, Peking duck, have also created a unique Quanjude 'All-duck Banquet'. Their business is growing by the day: they have over 100 branch restaurants spread throughout the country, and they are the first long-established restaurateurs to have been floated on the A-share domestic stock exchange.

Situated in the most flourishing district of Beijing—Wangfujing, Donglaishun Restaurant evolved from once a small Hui nationality porridge stall, to selling lamb hotpot, and eventually becoming the number one "old name" in this business. Aside from its authentic Beijing Lamb Hotpot, there came to be no less than 200 other qingzhen (Hui Muslim) dishes such as Mixed Fungus and Tremella, Roast Lamb Leg, Innards in Clear Soup, Hand-served Lamb, Fired-roasted Lamb Tail and so on. Its snacks such as butter fried cake, and walnut cream are also quite famous. Walking into Donglaishun, tasting all kinds of qingzhen foods and Beijing's Lamb Hotpot, one could obtain from it a sense of contentment.

Compared to home-style cooking, regional cuisines and the "old name" restaurants, Chinese fast food, adopting foreign business formulas, have only been around for several decades, but have spread to every corner of cities of all sizes. These types of restaurants were revealed to the public in all new modes of management. With their long store hours and traditional tasting foods, fast food stores can meet customers' needs for meals at any time. Plus low pricing, comprehensive selections, numerous tastes, and clean individual servings allowed fast foods restaurants to develop with powerful momentum. With McDonald's, KFC, Pizza Hut and other Western fast food franchises taking hold in the country, Western fast food, characterized by coke, burgers and pizzas, are doing quite well in China despite their not-so-cheap prices.

In fact, real Western food first appeared in China long ago, more than 7 centuries ago to be exact, when the Italian traveler, Marco Polo visited China. Ways of making Western food were introduced to China. But these foods only appeared in overseas Chinese homes or Chinese imperial and noble families. There was a long way to go before Western foods become an integral part of the Chinese food service industry. From late 19th to early 20th century, accompanying foreign invasion from the Western superpowers, Chinese people who worked for foreigners mastered Western food-making techniques. Making and eating Western food for the Chinese was no longer a fancy practice. It even developed into a whole separate line of business.

In the twenty-some years following the implementation of China's open-up policy, number of restaurants selling specialty food from both East and West was on the rise. Some are "old name" Western food restaurants located inside high-class hotels; others were newly opened independent restaurants at places where foreigners gather to work and live. Some cities with tourist attractions even built specialty food streets, providing various foreign foods for foreign tourists, complete with dining and leisure alternatives. The foreign foods offered by many restaurants could each form their own line of business, with their unique Chinese interpretation for different regions, cultures and customs of different countries, enriching the daily lives of the Chinese and stimulated the growth towards prosperity of the Chinese food service industry. People from different countries joyfully experience respect and tolerance for one another in food.

< A restaurant with traditional decorations (Photo by Feng Gao, provided by Imagechina)

Appendix: Chronological Table of the Chinese Dynasties

The Paleolithic Period	Approx. 1,700,000–10,000 years ago
The Neolithic Age	Approx. 10,000–4,000 years ago
Xia Dynasty	2070–1600 BC
Shang Dynasty	1600–1046 BC
Western Zhou Dynasty	1046–771 BC
Spring and Autumn Period	770–476 BC
Warring States Period	475–221 BC
Qin Dynasty	221–206 BC
Western Han Dynasty	206 BC–AD 25
Eastern Han Dynasty	25–220
Three Kingdoms	220–280
Western Jin Dynasty	265–317
Eastern Jin Dynasty	317–420
Northern and Southern Dynasties	420–589
Sui Dynasty	581–618
Tang Dynasty	618–907
Five Dynasties	907–960
Northern Song Dynasty	960–1127
Southern Song Dynasty	1127–1279
Yuan Dynasty	1206–1368
Ming Dynasty	1368–1644
Qing Dynasty	1616–1911
Republic of China	1912–1949
People's Republic of China	Founded in 1949